# Lecture Notes in Computer Science 9303

Commenced Publication in 1973
Founding and Former Series Editors:
Gerhard Goos, Juris Hartmanis, and Jan van Leeuwen

More information about this series at http://www.springer.com/series/7407

Michael Lones · Andy Tyrrell
Stephen Smith · Gary Fogel (Eds.)

# Information Processing in Cells and Tissues

10th International Conference, IPCAT 2015
San Diego, CA, USA, September 14–16, 2015
Proceedings

 Springer

*Editors*

Michael Lones
Heriot-Watt University
Edinburgh
UK

Andy Tyrrell
University of York
York
UK

Stephen Smith
University of York
York
UK

Gary Fogel
Natural Selection, Inc.
San Diego, CA
USA

ISSN 0302-9743          ISSN 1611-3349   (electronic)
Lecture Notes in Computer Science
ISBN 978-3-319-23107-5      ISBN 978-3-319-23108-2   (eBook)
DOI 10.1007/978-3-319-23108-2

Library of Congress Control Number: 2015946101

LNCS Sublibrary: SL1 – Theoretical Computer Science and General Issues

Springer Cham Heidelberg New York Dordrecht London

Springer International Publishing AG Switzerland is part of Springer Science+Business Media
(www.springer.com)

# Preface

Celebrating the 20th anniversary of the IPCAT series, the 10th International Conference on Information Processing in Cells and Tissues took place during September 14–16, 2015, at the Embassy Suites, downtown San Diego, CA. The IPCAT series of conferences began in Liverpool in 1995 as a venue to bring together multidisciplinary scientists interested in modelling the processes that take place within biological cells and tissues. This was followed by events held in Sheffield, Indianapolis, Brussels, Lausanne, York, Oxford, Ascona, and Cambridge. Over the years, the conference has been organized by biologists, mathematicians, computer scientists, and electronic engineers, but has always aimed to attract a diverse and multidisciplinary group of delegates. As noted by Ray Paton and Mike Holcombe in the foreword to the first IPCAT workshop, "One of the key motivations underlying the first IPCAT Workshop was to attempt to provide a common ground for dialogue and reporting research without emphasising one particular research constituency or way of modelling or singular issue in this area."

For IPCAT 2015, we addressed the diversity of the IPCAT audience by assembling Organizing and Program Committees comprising people with backgrounds in biology, medicine, mathematics, computer science, the natural sciences, and engineering. To reflect the differing publication norms in diverse fields, we gave authors the option of submitting either an extended abstract or a full paper, treating these equally during the review and ranking process. To complement the technical program, we invited four renowned scientists to give keynote presentations. These each addressed particular aspects of information processing in biological cells and tissues:

- Lee Altenberg (Konrad Lorenz Institute, Austria), "How Might Evolutionary Theory Inform Research on Information Processing in Cells and Tissues?"
- Kwang-Hyun Cho (KAIST, South Korea), "Unraveling the Information Processing Machinery Within a Living Cell"
- Terry Gaasterland (UCSD, USA), "Genome Variation in Regulatory Regions and Impact on Human Diseases"
- Marco Salemi (University of Florida, USA), "Phylodynamic Analysis of Viral and Bacterial Pathogens in the Genomics Era"

We would like to thank all the people involved in the organization and realization of IPCAT 2015, especially the authors, the invited speakers, and the members of the Program Committee, whose time and effort were central to the conference's success. We would also like to take this opportunity to remember Ray Paton, whose dedication and enthusiasm were central to the success of the IPCAT series.

September 2015

Michael Lones
Andy Tyrrell
Stephen Smith
Gary Fogel

# Organization

## Organizing Committee

### General Chairs

Andy M. Tyrrell      University of York, UK
Stephen L. Smith      University of York, UK

### Program Chair

Michael A. Lones      Heriot-Watt University, UK

### Local Chair

Gary Fogel      Natural Selection Inc., USA

## Program Committee

| | |
|---|---|
| Alexander Bockmayr | Freie Universität Berlin, Germany |
| Rachel Cavill | Maastricht University, The Netherlands |
| Jerry Chandler | George Mason University, USA |
| Cristina Costa Santini | King Saud University, Saudi Arabia |
| Ron Cottam | Vrije Universiteit Brussel, Belgium |
| Peter Erdi | Kalamazoo College, USA |
| Luis Fuente | Oxford Brookes University, UK |
| Jean-Louis Giavitto | IRCAM, France |
| Pauline Haddow | Norwegian University of Science and Technology, Norway |
| David Halliday | University of York, UK |
| Arun Holden | University of Leeds, UK |
| Sam Marguerat | Clinical Sciences MRC, UK |
| Maizura Mokhtar | University of Sheffield, UK |
| J. Manuel Moreno | Universitat Politècnica de Catalunya, Spain |
| Chrystopher L. Nehaniv | University of Hertfordshire, UK |
| Simon O'Keefe | University of York, UK |
| Hiroshi Okamoto | RIKEN Brain Science Institute, Japan |
| Tjeerd Olde Scheper | Oxford Brookes University, UK |
| Leslie Smith | University of Stirling, UK |
| Christof Teuscher | Portland State University, USA |
| Jim Tørresen | University of Oslo, Norway |
| Alexander Turner | University of York, UK |
| Juanma Vaquerizas | Max Plank Institute for Molecular Biomedicine, Germany |

# Contents

## Biochemical Regulatory Networks

## Metabolomics and Phenotypes

## Neural Modelling and Neural Networks

# Biochemical Information Processing

# Surface-Immobilised DNA Molecular Machines for Information Processing

Katherine E. Dunn[(✉)], Tamara L. Morgan, Martin A. Trefzer,
Steven D. Johnson, and Andy M. Tyrrell

Department of Electronics, University of York, York, UK
katherine.dunn@york.ac.uk

**Abstract.** The microscopic information processing machinery of biological cells provides inspiration for the field of molecular computation, and for the use of synthetic DNA to store and process information and instructions. A single microlitre of solution can contain billions of distinct DNA sequences and consequently DNA computation offers huge potential for parallel processing. However, conventional data readout systems are complex, and the methods used are not well-suited for combination with mainstream computer circuits. Immobilisation of DNA machines on surfaces may allow integration of molecular devices with traditional electronics, facilitating data readout and enabling low-power massively parallel processing. Here we outline a general framework for hybrid bioelectronic systems and proceed to describe the results of our preliminary experiments on dynamic DNA structures immobilised on a surface, performed using QCM-D (quartz crystal microbalance with dissipation monitoring), which involves the use of acoustic waves to probe a molecular layer on a gold-coated quartz sensor.

**Keywords:** DNA computation · Surface-immobilisation · Quartz crystal microbalance with dissipation monitoring · DNA strand displacement

## 1 Information Processing with Bioelectronic Systems

The Central Dogma of molecular biology [4] describes how information is stored and processed in living cells. According to this scheme, data is encoded in biological polymers (nucleic acids and proteins) and processed by specialised molecular machines, and the Dogma can provide inspiration for the development of a framework for hybrid bioelectronic computational systems. We present such a framework in Fig. 1. In our scheme there are three nodes for information processing, which we can regard as phases.

In our framework, each phase contains machinery which can act as memory for storage of data or instructions, or as processors which can execute programs in response to inputs received from the other phases or externally. Information can be transmitted between the three phases or exported from the system as readout. Our framework does not exclude any conceivable links between phases,

© Springer International Publishing Switzerland 2015
M. Lones et al. (Eds.): IPCAT 2015, LNCS 9303, pp. 3–12, 2015.
DOI: 10.1007/978-3-319-23108-2_1

although some connections would be easier than others to implement in a real-world system. Information can also be exchanged between different machines contained within the same phase. It is important to note that our framework is not prescriptive and could readily be expanded to accommodate technological developments, while phenomena not explicitly mentioned here could be harnessed for information processing within this scheme.

Information transfer between any two phases in the framework is symmetrical - each link between phases is bidirectional, although the mechanism, energetics and kinetics for information transfer is not generally the same when the direction is reversed. Transmission of information from one phase to another is a read/write process, in which one phase writes information to the next or the downstream phase reads information from the previous one. Within each phase, information is processed through a standard cycle of *fetch*, *decode* and *execute* (not shown explicitly in the diagram).

The solid state phase (Fig. 1) represents systems based on conventional semiconductor technology or other solid state devices. This could include photonic components, which have considerable potential for computation [12] but are presently at an early stage of development. Light may also be used for information transfer processes. For instance, many chemical reactions can be activated by light [13] and optical fibres can be used for high-speed data transfer. Initially, the solid state phase of hybrid bioelectronic systems is likely to comprise conventional silicon-based devices due to the rapidity of operations, coupled with the extreme maturity of the underlying technology and manufacturing techniques.

As we envisage it, the surface phase comprises discrete surface-immobilised molecular machines, which encode and store information in the primary sequence of the biological polymers from which they are made. These molecular machines execute programs by undergoing a change in state (e.g. through change in exposed DNA sequence) or conformation. In our scheme, the solution phase contains similar molecular machines [2] which float freely in an aqueous environment. They perform operations and store information in the same way as their surface-immobilised counterparts, but their behavior may differ due to the absence of surface-specific effects such as localised electric fields or molecular crowding.

In both the surface and solution phases, exchange of data and instructions between different machines in the same phase is accomplished by means of the intermolecular interactions that regulate self-assembly of DNA and protein macromolecules, i.e. van der Waals interactions, electrostatic forces, entropic or steric effects, and covalent bonding. Similarly, information is conveyed from solution to surface or vice versa by means of molecular interactions which drive binding or dissociation of specific components within a molecular machine. For instance, a molecular species released as the output from a solution phase process could be programmed to bind to a surface-immobilised molecular machine as an input. Information transfer to the solid phase from surface-immobilised machines or their solution-phase equivalents would typically be accomplished by charge transfer, for example the direct transport of electrons between redox-active

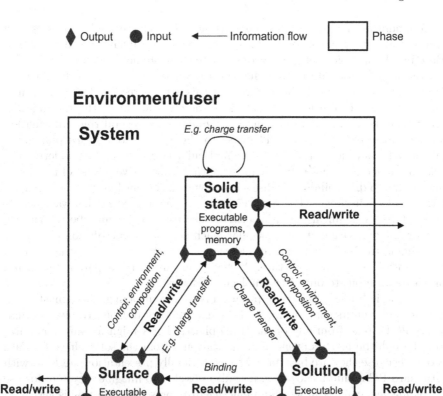

**Fig. 1.** General framework for a hybrid bioelectronic system in which molecular devices are integrated with solid-state electronics. There are three phases, between which information can be exchanged. In each phase, programs can be executed and information stored in memory. The solid-state phase may be represented by conventional silicon technology or any suitable alternative. The surface phase consists of immobilised molecular machines, and the solution phase contains all relevant molecules which are not bound to the surface. Arrows indicate transfer of information between phases, or the operations of reading/writing data. Outputs and inputs are represented by the indicated symbols.

molecules (such as methylene blue, ferrocene or Nile Blue) and the underlying substrate. This approach to electronic transduction of molecular binding is well-established following recent advances in hybrid DNA-electronic clinical sensors [11].

Communication with the solid phase could also occur through processes such as mass transfer but a change in charge distribution or dynamics is most appropriate in a hybrid bio*electronic* system where semiconductor devices are used.

Transfer of information from the solution phase to the solid state must by definition proceed through the surface, but this is to be regarded as a direct and passive link if the solution-solid interface merely transduces the information, rather than modifying or processing it *en route*. As a general rule, information can be transferred from the solid phase to the solution or surface phases via control systems which can regulate the molecular interactions that underpin the molecular machines. This can be achieved either by selective release of molecular components to drive binding or dissociation within the machinery or via changes of the local environment. Factors which can be modulated in this way include the local electrochemical potential (via an electrode, which may be the surface used for immobilisation of the molecular machines), the local solution pH (either by changing the buffer with microfluidics or electrochemically [7]), the local ion concentration or the temperature. A change in any of these parameters could constitute an input to one of the molecular phases.

If a generic framework for computation is to have meaning, there must be an interface between its components and the external world. Conceivably, information could be read from any of the three phases. For straightforward computation, it would be most convenient to use outputs from the solid phase for data readout because they are best suited to user-friendly presentation, perhaps with a standard PC monitor. However, it is also possible to imagine a system for use in a biomedical context in which an inactive drug present in solution is activated as result of a computation carried out jointly by all three phases, and the output would then consist of the activated drug, as released into the bloodstream. Alternatively, for some applications the desired output might be a visible change on the surface, such as a change of color or regulation of fluorescence intensity. Information could be supplied as input to any phase by programming the solid state phase, by pumping a new species of molecule into the solution, or by changing the properties of the surface. Significantly, bioelectronic computers could be interfaced with systems which would not be accessible to conventional devices.

With the hybrid bioelectronic systems described by our framework, we can easily move away from the paradigm of binary logic. While analogue electronic computation does exist, most conventional computer systems are based on digital logic. In contrast, the behavior and response of a molecule or molecular ensemble is inherently not binary. In a hybrid bioelectronic system it would therefore be possible to express information through the amplitude and temporal response of a signal rather than using a simple binary encoding. Furthermore, the intrinsic properties of molecular machines make them extremely well-suited for parallel processing due to the enormous potential diversity of molecular species present in the reaction mixture or on the surface. It is, however, important to note that not all possible species are usable in practice, because the conformation of some individuals may be restrictive or interactions between some species may be undesirable. Parallel processing can also be implemented *in silico* using specialised architectures and algorithms.

The framework we have developed suggests that surface immobilisation of molecular devices is critical if they are to be integrated with conventional solid state electronic systems for information processing purposes. For such applications, information must be encoded within the structure of the molecule, and if the approach is to be scalable the structure must permit expression of a range of variants, to represent different instructions or items of data. The molecule must also be able to switch from one configuration to another in order to execute an instruction, and for practicality it must be relatively stable and economical to synthesise. DNA meets all of these criteria, and its attributes make it a particularly attractive candidate for synthesis of integrated systems.

It is possible to design and synthesise DNA strands which will interact in a highly predictable manner, and the density of information stored in DNA is huge - approximately $2.5 \times 10^{26}$ bytes m$^{-3}$, in comparison to around $1.6 \times 10^{16}$ bytes m$^{-3}$ for a typical magnetic hard disk drive[1]. DNA could be used for data storage, either for long term memory [8], as in living organisms, or for recording temporary instructions. DNA devices can switch from one state to another in response to a supplied input, which effectively corresponds to processing an instruction - the information encoded in the sequence of the toehold specifies how the incoming strand should interact with the constructs on the surface, and hence determines the response of the molecular machinery. Manipulation of the dynamics of this process will allow the molecular switching process to be controlled precisely, enabling implementation of molecular circuits which absorb and process inputs according to a designed scheme, in order to perform logical operations and potentially complex computations.

The use of DNA for computation was pioneered by Adleman, who demonstrated in 1994 that the standard tools of molecular biology could be used to solve an instance of the travelling salesman problem [1]. More recently, DNA has been used to build a finite state machine [3], a neural network [10] and logic-gated nanorobots [6], among other systems. Most of these devices rely for their operation on the phenomenon of toehold-mediated strand displacement [14]. This is the process whereby an incumbent DNA strand is displaced from a double-stranded molecule by an invading strand, which binds transiently to a single-stranded toehold, as shown in Fig. 2a. Most previously reported DNA nanodevices operate in solution, but immobilising such machines on a surface will ultimately provide opportunities to create an interface between the solid state and biomolecules in solution. It will also be possible to create cascades of devices, consisting of multiple machines which operate sequentially, enabling construction of complex molecular logic circuitry.

---

[1] In DNA, a single base pair occupies $1\,\mathrm{nm}^3$ and stores 2 bits of information; for the magnetic hard disk, capacity is assumed to be 1 TB, radius 4.5 cm and depth 1 cm.

## 2  Experimental Studies of DNA Machines on Surfaces

Surface-immobilised DNA devices may be studied with the technique of QCM-D (quartz crystal microbalance with dissipation monitoring) [5]. This involves the use of a thin circular sensor consisting of a quartz crystal with a gold electrode on each side (Fig. 2b). A voltage is applied across the crystal, causing it to oscillate at its resonant frequency. The resonant frequencies shift when molecules are deposited on the sensor, and the size of the shift ($\Delta f$) depends on the mass adsorbed, which means that the crystal acts as a microbalance or mass sensor.

When the voltage is switched off the oscillation decays, at a rate which depends on the viscoelasticity of the molecular layer and the conformation of its constituent molecules [9]. This is quantified by the dissipation, $D$, which is defined mathematically as the inverse of the quality factor of the oscillator. If the molecular layer is highly viscoelastic, $D$ will be large because the energy will be transferred easily from the crystal to the surrounding solution.

As the acoustic wave propagates into the solution from the crystal, the size of the oscillation decreases. The amplitude decays exponentially with distance from the surface, and for the sensors used here the penetration depth in pure water is approximately 250 nm for the fundamental frequency (5 MHz). The higher resonant frequencies probe an even smaller region, and QCM-D is therefore very sensitive to processes occurring at or near surfaces. The combination of its sensitivity with its capacity to monitor dynamic processes in real-time and its ability to directly probe molecular conformations makes it ideally suited for our studies of surface-immobilised DNA molecular machines for information processing.

### 2.1  Materials and Methods

QCM-D experiments were performed using a Q-Sense E4 system with gold-coated QSX 301 quartz sensors, where the apparatus and sensors were supplied by Biolin Scientific. According to the manufacturers, the maximum mass sensitivity of the system is $\sim 0.5\,\mathrm{ng\,cm^{-2}}$ in liquid, and a more typical value would be $\sim 1.8\,\mathrm{ng\,cm^{-2}}$. The area of a sensor is approximately $1.5\,\mathrm{cm^2}$.

Sensors were cleaned before use as follows: 10 min UV-ozone treatment; sonication for 10 min in (1) a 2 % solution of Hellmanex III, (2) ultrapure (MilliQ) water (twice); drying with $N_2$ gas; 30 min UV-ozone treatment; soaking in 100 % ethanol for at least 30 min; drying with $N_2$ gas. The concentration of DNA used for immobilisation was typically 300 nM and for backfilling the concentration of mercaptohexanol was 1 mM.

The sequence of the thiolated DNA strand was 5'-ACACGCATACACCCAT-(thiol)-3' and the sequence of the extended strand which bound to it was 5'-ATGGGTGTATGCGTGT**TTAAAGACCCTAAGCT**-3'. The thiol was provided in oxidised form as a disulphide with a short alkyl chain on each side. Chemicals were acquired from Sigma Aldrich and DNA was purchased from Integrated DNA Technologies (IDT), with HPLC purification for thiol-modified strands. Oligos were stored at 4°C in 1×TE.

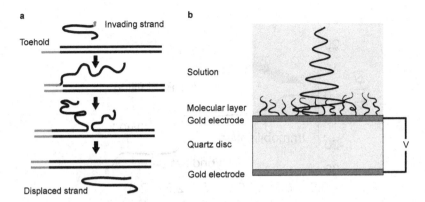

**Fig. 2.** (a) The process of toehold-mediated DNA strand displacement. The invading strand binds to the single-stranded overhang ('toehold') of the original duplex and stochastically displaces the incumbent strand from the complex via branch migration. It is possible for the invading strand to dissociate before branch migration occurs, and the most stable configuration of strands is that with the greatest number of base pairs. (b) The principle of QCM-D. The voltage applied causes the piezoelectric quartz disc to oscillate, and acoustic waves propagate from the sensor into solution, decaying exponentially with distance from the surface. The resonant frequencies of the crystal and the dissipation of energy from the acoustic waves depend on the mass and conformation of the molecules adsorbed on the sensor.

## 2.2 Experimental Results

Figure 3 shows the frequency and dissipation changes observed during immobilisation of a single-stranded DNA oligonucleotide with a thiol modification. As molecules bound to the surface, the mass of the layer on the sensor increased, and the frequency decreased. At each stage, the dissipation increased, which indicates that energy was transferred more easily from sensor to solution as more molecules adsorbed. Immobilisation was a comparatively slow process, occurring over a timescale of approximately 30–40 min. After immobilisation, we applied a backfilling agent (mercaptohexanol, MCH) to reduce non-specific adsorption and improve the quality of the DNA layer. In the final stage of the experiment we added the reverse complement of the immobilised strand and we observed that hybridisation occurred rapidly, within a few minutes (Fig. 3).

We examined a range of immobilisation conditions, and we found that the concentration of thiolated DNA only weakly affects the amount of DNA which binds to the surface. We established that the presence of salt is essential for immobilisation, for charge screening, and confirmed that it is essential in all experiments to establish baseline measurements for frequency and dissipation in the correct buffer because these are strongly affected by salt and other buffer components. We observed that immobilisation of DNA duplexes tends to proceed more rapidly than that of single-stranded oligonucleotides, presumably because double-stranded molecules are more rigid and thus self-organise more easily.

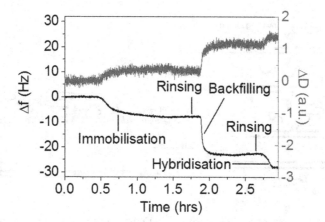

**Fig. 3.** (a) Frequency and dissipation changes measured using QCM-D during immobilisation of single-stranded DNA oligonucleotides, backfilling with mercaptohexanol, and hybridisation with the reverse complement of the immobilised strand.

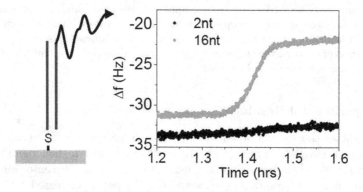

**Fig. 4.** Sketch of experimental setup for strand displacement measurements, and frequency changes associated with strand displacement for toeholds of length 2 (black) or 16 (grey) nucleotides. The concentration of the displacing strand was 600 nM in each case. DNA was immobilised on the electrode via a single thiol-gold bond and experiments were performed at or slightly below room temperature, in TE buffer with 1M NaCl, at a flow rate of $20 \, \mu$L min$^{-1}$. Data is shown for the 13$^{th}$ resonance; this has the shortest penetration depth and is therefore the most sensitive to the surface.

For our investigations into strand displacement we used a pre-formed duplex, consisting of a 16 base-pair double-stranded region and a 16 nucleotide single-stranded overhang at the end furthest from the surface, as sketched in Fig. 4. The process of strand displacement generated a double-stranded waste product which remained in solution, while the single-stranded molecule continued to be attached to the surface. The mass of the molecular layer on the surface therefore decreased as displacement proceeded, and the frequency increased (Fig. 4).

Our data (Fig. 4) shows that toehold-mediated DNA strand displacement occurs more slowly at the surface than in solution. In our experiments, the fastest reactions occurred within minutes, with a rate constant of the order of $k = 10^3 \, \mathrm{M^{-1}s^{-1}}$. In contrast, reactions in solution may have rate constants of around $k = 10^6 \, \mathrm{M^{-1}s^{-1}}$, giving a characteristic timescale of the order of seconds. This huge difference arises because an immobilised duplex has fewer degrees of freedom than a molecule which floats freely in solution, and the availability of the toehold is effectively restricted, such that only the invading strands which approach the surface have the opportunity to initiate displacement. Interactions between duplexes within the surface-assembled monolayer may also hinder access to the toeholds. We observed minimal strand displacement for the shortest toeholds (2–3 nucleotides), and we found that displacement occurred rapidly for longer toeholds, as in solution [14].

## 3  Conclusion

In order to fully harness the information processing power of DNA molecular machines on surfaces, a full understanding of the dynamics of surface-immobilised DNA machines is required, and our investigations show that QCM-D has great potential in this area. The results of our experiments reveal that DNA toehold-mediated strand displacement on surfaces follows some of the same trends as in solution, but some features of the process are different. Displacement occurs comparatively slowly in immobilised constructs, and this will need to be taken into account if processes carried out by surface-based machinery are to be co-ordinated with those occurring in other phases of a hybrid system (Fig. 1). Most dynamic DNA machines rely on the phenomenon of strand displacement and it will therefore be necessary to understand how to control the dynamics of this process in order to maximise the performance of the immobilised machines comprising the surface phase of our framework.

Further work will shed light on other aspects of the behavior of surface-immobilised DNA machinery, and may demonstrate whether other DNA structures and alternative molecules have potential for information processing. For instance, the use of DNA constructs with secondary structure motifs may allow the availability of selected domains to be controlled, which could facilitate the design of logic circuits. In addition, incorporation of DNA structures with alternative geometries ($i$−motif, G-quadruplex etc.) could enable switching events to be defined by a conformational change, occurring in response to a shift in environmental properties, which would enable sensing of external conditions and processing of different forms of information. In principle, RNA strands could be designed and used in the same way as those of DNA, but RNA is more difficult to work with and considerably more expensive. Peptides and proteins would present more complex behavior than nucleic acids and might therefore be more versatile, but it would be more difficult to predict the nature of the intermolecular interactions, which would complicate efforts to design reliable systems.

In conclusion, the results obtained in the experiments described here will underpin efforts to design surface-immobilised DNA machinery, thus opening the door to integration of such molecular devices with more conventional computational systems for information processing applications, in accordance with the general framework we presented in Fig. 1.

**Acknowledgements.** The authors would like to thank the EPSRC for their support of this work through the Platform Grant EP/K040820/1, and we also wish to thank the University of York for the award of an Institutional Equipment Grant. Data created during this research is available at the following DOI: 10.15124/3861217c-b93a-4df9-b03a-2e102e5b47c7.

# References

1. Adleman, L.: Molecular computation of solutions to combinatorial problems. Science **266**(5187), 1021–1024 (1994)
2. Bath, J., Turberfield, A.: DNA nanomachines. Nat. Nanotechnol. **2**, 275–284 (2007)
3. Costa Santini, C., Bath, J., Tyrrell, A., Turberfield, A.: A clocked finite state machine built from DNA. Chem. Commun. **49**, 237–239 (2013)
4. Crick, F.: Central dogma of molecular biology. Nature **227**, 561–563 (1970)
5. Dixon, M.: Quartz crystal microbalance with dissipation monitoring: enabling real-time characterization of biological materials and their interactions. J. Biomol. Tech. **19**, 151–158 (2008)
6. Douglas, S., Bachelet, I., Church, G.M.: A logic-gated nanorobot for targeted transport of molecular payloads. Science **335**, 831–834 (2012)
7. Gabi, M., Sannomiya, T., Larmagnac, A., Puttaswamy, M., Voros, J.: Influence of applied currents on the viability of cells close to microelectrodes. Integr. Biol. **1**, 108–115 (2009)
8. Goldman, N., Bertone, P., Chen, S., Dessimoz, C., LeProust, E., Sipos, B., Birney, E.: Towards practical, high-capacity, low-maintenance information storage in synthesized DNA. Nature **494**, 77–80 (2013)
9. Papadakis, G., Tsortos, A., Bender, F., Ferapontova, E., Gizeli, E.: Direct detection of DNA conformation in hybridization processes. Anal. Chem. **84**(4), 1854–1861 (2012)
10. Qian, L., Winfree, E., Bruck, J.: Neural network computation with DNA strand displacement cascades. Nature **475**, 368–372 (2011)
11. Strehlitz, B., Nikolaus, N., Stoltenburg, R.: Protein detection with aptamer biosensors. Sensors **8**, 4296–4307 (2008)
12. Thompson, M., Politi, A., Matthews, J., O'Brien, J.: Integrated waveguides for optical quantum computing. IET Circuits Devices Syst. **5**, 94–102 (2011)
13. Yang, Y., Endo, M., Hidaka, K., Sugiyama, H.: Photo-controllable DNA origami nanostructures assembling into predesigned multiorientational patterns. JACS **134**, 20645–20653 (2012)
14. Zhang, D., Winfree, E.: Control of DNA strand displacement kinetics using toehold exchange. J. Am. Chem. Soc. **131**(47), 17303–17314 (2009)

# Scalable Design of Logic Circuits
# Using an Active Molecular Spider System

Dandan Mo[1]([✉]), Matthew R. Lakin[1,2,3], and Darko Stefanovic[1,3]

[1] Department of Computer Science, University of New Mexico,
Albuquerque, NM, USA
{mdd,mlakin,darko}@cs.unm.edu

[2] Department of Chemical and Biological Engineering, University of New Mexico,
Albuquerque, NM, USA

[3] Center for Biomedical Engineering, University of New Mexico,
Albuquerque, NM, USA

**Abstract.** As spatial locality leads to advantages of computation speed-up and sequence reuse in molecular computing, molecular walkers that exhibit localized reactions are of interest for implementing logic computations. We use molecular spiders, which are a type of molecular walkers, to implement logic circuits. We develop an extended multi-spider model with a dynamic environment where signal transmission is triggered locally, and use this model to implement three basic gates (AND, OR, NOT) and a mechanism to cascade the gates. We use a kinetic Monte Carlo algorithm to simulate gate computations, and we analyze circuit complexity: our design scales linearly with formula size and has a logarithmic time complexity.

**Keywords:** Molecular spiders · Logic circuits · Parallel evaluation · Localized signal transmission

## 1 Introduction

Molecular walkers are synthetic molecular machines inspired by natural biological motors. Previous studies [4,7,9,11,13] have shown that walkers can move directionally and autonomously on a pre-programmed track via localized reactions. Spatial locality can overcome the challenges of computation speed-up and sequence reuse in molecular computing where all the reactants diffuse freely in a mixed solution [2,5]. Hence, a walker system with inherent spatial locality has potential to perform more complex computational tasks. We investigate the computational power of a walker system by using it to implement scalable logic circuits.

We consider a molecular *spider* system, where a spider is a type of molecular walker. Molecular spiders [1,11,12] with varying number of legs move stochastically on a surface formed by sites containing DNA segments, and present biased behaviors due to different reactions with fresh sites (catalytic cleavage) and visited sites (dissociation). We extend previous models [1,11,12] to implement three basic logic gates (AND, OR, NOT), and cascade the gates to construct

© Springer International Publishing Switzerland 2015
M. Lones et al. (Eds.): IPCAT 2015, LNCS 9303, pp. 13–28, 2015.
DOI: 10.1007/978-3-319-23108-2_2

logic circuits. We use multiple spiders in the model, and we assume spiders behave unbiasedly with equal transition rates to all sites. Sites are divided into *normal sites* that are non-alterable and *functional sites* that can be altered via catalytic cleavage and/or strand displacement. We can encode signals into functional sites. Signal transmission [5,6] is triggered locally when a spider interacts with a signal-carrying site, which dynamically changes the state of the spider or of the environment. We call this extended system an *active* molecular spider system.

In our design, each variable is represented by a moving spider that has two legs and one arm. The arm has two possible states, 0 or 1, representing the boolean value of the spider. Each gate is represented by a layout of different sites on a 2D lattice. In a single gate, spiders with different values will take different paths from their input locations. We arrange different functional sites on different paths, such that only the spider with the correct computation result will be directed to the output location via interactions between spiders and functional sites. On reaching the output location, a spider reports the computation result, and we call this spider the *reporting* spider. We cascade logic gates by connecting them such that only the reporting spider leaves the upstream gate and enters the downstream gate. We design a mechanism for *exit* from gates to implement gate cascades that allow parallel evaluation. As an example, Fig. 7 shows a logic circuit where input spiders $X$ and $Y$ are initially placed at the input locations of two NOT gates, and the NOT gates are connected to the same AND gate via *exit* mechanisms. Spiders move within the circuit, and the spider reaching the output location reports the computation result.

Molecular circuits using *DNA Strand Displacement* [8] in a well-mixed solution use relatively high and low concentration of a species to represent Boolean values 1 and 0, or use two separate species in a dual-rail encoding. Here, we use spiders with arm state 1 or 0 to represent Boolean values, thus we remove potential ambiguity from result reporting. Since Boolean values are carried by spiders moving from an upstream gate to a connected downstream gate, all gates are designed individually, thus, modularity is ensured. Previous work on tethered circuits [2,5] also ensures modularity and unambiguity, but takes a different approach where Boolean values are represented by the existence of a sequence. Modularity is ensured by spatially isolating different gates on a surface, e.g., a DNA origami tile such that only gates in close proximity can interact with each other.

Previous work [3] has used a walker system to construct logic circuits with spatial locality, but it lacks modularity and is limited to sequential evaluation due to its design where the circuit constructed is in the form of a *Binary Decision Diagram* (BDD). A walker initially placed at the root node walks along a path unblocked by externally-added strands to reach a leaf node representing True or False, causing a fluorophore change to report the computation result. For practical reasons, this reporting strategy needs two parallel circuits that detect fluorophore change at the True nodes and False nodes respectively to avoid ambiguity. Our design uses the reporting spider to avoid reporting problems [3], and we support parallel evaluation. As a result, to evaluate an $m$-clause 3-CNF circuit, we need time $O(\log m)$ while the circuit [3] needs time $O(m)$. We use the same linear space complexity $O(m)$ as in the circuit [3], and it is easier to construct large circuits using our design because of its modularity.

Using an extended *active* multi-spider system, while keeping the advantages related to spatial locality, our design ensures modularity, unambiguity, and scalability. We will describe the model in Sect. 2, and introduce how to construct the logic circuits in Sect. 3 with simulation results and complexity analysis. A formal definition of the model is given in Sect. 4. We give conclusions and discuss current challenges and future work in Sect. 5.

## 2   Model Description

Our long-term goal is to realize the circuits we describe here with a physical implementation based on molecular spiders [4,7]. Therefore, our model draws from the existing models of molecular spiders [9,11] and extends them to describe the richer functionalities of the walkers we hope to build. In spite of these extensions, we will use the evocative term "spider" throughout the paper.

A molecular spider has a body and three limbs, two legs and an "arm", which it can use to attach to chemical sites on a surface. There is exclusion: at most one limb can be attached to a given site at a time. Different types of sites are laid out on a square lattice, $\mathbb{Z}^2$. A set of contiguous sites can form a *track* on which the spiders can move.

We model a spider's body as a single point, and the limbs as having equal length. This leads to the following postulated "hand-over-hand" gait [9]: at any given time, exactly two limbs are attached to the surface, and they are attached to nearest-neighbor sites. We call the sites a limb has bound to the *attachment points*. There are always two attachment points for each spider, and they are adjacent to each other. A moving step occurs when a spider detaches one of its limbs from an attachment point $p \in \mathbb{Z}^2$, and attaches to a site $p' \in \mathbb{Z}^2$. Figure 1 shows a transition step of a spider where there are four *reachable* sites that the spider can potentially transit to. However, a spider might not attach to a reachable site because whether a reachable site is *available* depends on the state of the site and of the limb, which will be discussed in Sects. 3 and 4. When multiple spiders are moving on the track, one spider cannot attach to a site occupied by another spider.

Spiders move stochastically on the track, interacting with the normal sites. If they attach to functional sites, signal transmission is triggered locally between two adjacent sites, or between a site and the spider attached to it. Changes to the sites and spiders may happen during a step, which is crucial in the construction of a logic circuit. In the next section, we will explain how to use different sites to construct three basic gates (AND, OR, NOT) and cascade them to construct a logic circuit.

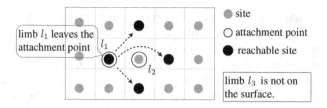

**Fig. 1.** A spider has limb $l_1$ and limb $l_2$ attached to the surface. When limb $l_1$ detaches from the left attachment point, four sites represented by the black dots are reachable for limbs $l_1$ and $l_3$. The arrows show the transitions of a spider to other sites via *hand-over-hand* movement.

## 3    Logic Circuit Construction

Each spider represents a Boolean variable. The value of the spider is indicated by its arm state, which is either 0 or 1. A logic circuit is formed by cascades comprising the basic logic gates (AND, OR, NOT). This combination of logic gates is complete for Boolean logic. A logic gate is an arrangement of different sites on a square lattice, including an output location and input locations. When spiders begin moving from the input locations, their interactions with the sites lead to changes to the sites and the spider values, which ends with one spider reaching the output location, and the value of this spider represents the computation result of the logic circuit. In this paper we do not concern ourselves with the issues of placement and routing of circuits in the plane, which are well studied in electronic circuit design.

### 3.1    Normal Sites and Functional Sites

We define the set of site types as $S = S_{norm} \cup S_{fun}$, where the *normal* sites $S_{norm} = \{s_l, s_1, s_0\}$ are non-alterable and the *functional* sites in $S_{fun}$ are alterable. A normal site of type $s_l$ binds to a spider's leg, and is used for the "wires" of a logic circuit. Sites of type $s_0$ and $s_1$ bind to the spider's arm if it has type 0 or 1, respectively. Sites of type $s_0$ and $s_1$ are placed at the beginning of two separate paths that branch out from a junction, directing a spider with different values to different paths (Fig. 2).

The junction design is used in the constructions for all gate types. Each logic gate has a set of functional sites placed on the paths branching out from the junction. After the spiders take their own paths at the junction according to their values, they will encounter different functional sites. The interactions between the spiders and the functional sites cause changes to the spider and the sites, directing one spider to the output location, reporting the result of the gate computation.

Before going to the details of each gate, we first introduce some important features of functional sites. (1) A functional site has a state among {on, off, trapped}. The spider can bind to an "on"-state site, cannot bind to an "off"-state site, and

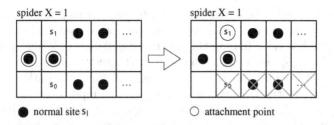

**Fig. 2.** If a spider has an arm type of 1, it binds to site $s_1$ at a junction. If a spider has an arm type of 0, it binds to site $s_0$ at a junction. Here a spider $X = 1$ follows the upper path by attaching to site $s_1$. It cannot follow the lower path.

cannot leave a "trapped"-state site by itself. (2) A functional site may or may not trap a spider. When it traps a spider, the site's state becomes "trapped". (3) A functional site may contain a signal of "turning on/off" or "switching to 1 or 0". The signal held in a functional site is sent out once it is attached by a spider. When a spider attaches to a site holding a signal, the signal "turning on/off" is sent to another site, setting its state "on" or "off"; the signal "switching to 1 or 0" is sent to the spider, changing its value to 1 or 0. When a functional site sends out its signal, it has no signal remaining. Signal transmission is allowed between a site and a spider that is attached to the site, or between two sites that are adjacent to each other. These features could be implemented via DNA strand displacement. We will discuss the AND and OR gate designs in Sect. 3.2 and the NOT gate design in Sect. 3.3.

### 3.2  Designs of the AND and OR Gates

We use three types of functional sites $s_t$, $s_p$, and $s_u$ in the designs of the AND gate and OR gate. Site $s_t$ can trap the spider attaching to it, so we place a site $s_t$ at the output location of the gate. The AND gate and OR gate each has two input spiders initially located at the two input locations, which are two junctions as shown in Fig. 2. Each input spider selects one of two possible paths when computation begins, where one path leads to the output location without any functional sites and the other path is merged into a crossroad in the middle of the lattice. We place an initially "off"-state site $s_p$ at the heart of the crossroad, which blocks the central path from the crossroad to the output location. We place a site $s_u$ adjacent to site $s_p$, which will send a "turning-on" signal to unblock site $s_p$ when a spider attaches to it, and trap that spider at the same time. The cooperation between sites $s_u$ and $s_p$ guarantees that only when both spiders meet at the crossroad can a spider take the central path to the output location.

Figure 3 shows the layout of the AND gate and OR gate. We explain how the AND gate works under four possible input assignments, and the OR gate follows a similar design. In the AND gate, the two input spiders $X$ and $Y$ are initially placed at two junctions as their input locations. When spiders $X$ and $Y$ are both 0, they both take the path starting with site $s_0$, which leads to the output location without any functional sites. In this case, whichever spider reaching the output location has value 0, reporting the result of $0 \wedge 0$ is 0. When spider $X = 0$ and spider $Y = 1$, spider $Y$ takes the path starting with site $s_1$, and gets stuck at the crossroad because site $s_p$ is "off". Spider $X$ takes the path starting with site $s_0$, and will eventually reach the output location, reporting the result of $0 \wedge 1$ is 0. When spider $X = 1$ and spider $Y = 0$, spider $X$ gets to the crossroad via the path starting with site $s_1$, and gets trapped at the crossroad due to the sites $s_t$ and $s_u$ placed on that path. Spider $Y$ is the only spider that can reach the output location in this case, reporting the result of $1 \wedge 0$ is 0. When both spiders are 1, they meet at the crossroad. Site $s_p$ is turned on by the signal sent from site $s_u$, so spider $Y$ can take the central path leading to the output location. Since spider $X$ is trapped at the crossroad, only spider $Y$ can reach the output location, reporting the result of $1 \wedge 1$ is 1.

Following a similar design, the layout of the OR gate is shown in Fig. 3. When both spiders are 0, they meet at the crossroad. Spider $X$ is trapped on sites $s_t$ and $s_u$, and spider $Y$ takes the unblocked central path to the output location, reporting the result of $0 \vee 0$ is 0. Under other input assignments, the 0-valued spider takes the path to the crossroad and gets stuck there, only the 1-valued spider can reach the output location, reporting the result of $1 \vee 0$, $0 \vee 1$, and $1 \vee 1$ is 1.

The movement of the spiders can be modeled as a continuous-time Markov process. We used a kinetic Monte Carlo algorithm to simulate gate computations. For each gate, under different assignments, we investigate the computation time using 10,000 iterations in each simulation. We assume the transition rate (the rate that a spider limb transits from one site to another) of each spiders is 1. Simulation results for the AND gate and OR gate are shown in Fig. 4. In the AND gate or OR gate, under a certain input assignment, the computation time follows a long-tailed distribution because spiders move stochastically. The computation time is the time spent on traversing the path taken by the reporting spider that reaches the output location; it is influenced by factors such as the transition rate or the length of the path. These factors have been discussed in previous work [10,11], so we do not focus on them in this paper.

### 3.3   Design of the NOT Gate

We use five types of functional sites in the NOT gate design. As is shown in the layout of the NOT gate in Fig. 5, site $s_t$ which can trap a spider that attaches to it is placed on the output location. Sites $s_{1\to0}, s_r^{I}, s_r^{II}$ and sites $s_{0\to1}, s_r^{I}, s_r^{II}$ form two different *switch* mechanisms $SW_{1\to0}$ and $SW_{0\to1}$ that are laid on two separate paths. The NOT gate has one input spider which is initially placed at a junction as the input location. Two separate paths branch out from the junction: one is taken by the 1-valued spider and contains mechanism $SW_{1\to0}$ that can change the spider value to 0, the other is taken by the 0-valued spider and contains mechanism $SW_{0\to1}$ that can change the spider value to 1. When a spider moves through a *switch* mechanism, its value is switched and its backward

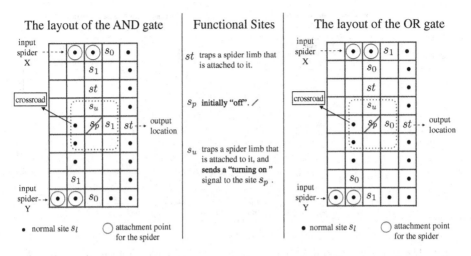

**Fig. 3.** The layout of the AND gate and OR gate. Three functional sites $s_t$, $s_p$, and $s_u$ used in the designs of these two gates are listed in the middle column. Normal site $s_1$ can only bind to an 1-valued spider and normal site $s_0$ can only bind to a 0-valued spider. In the AND gate, when both spiders are 1, they meet at the crossroad in the middle. Spider $X$ gets trapped at sites $s_t$ and $s_u$, site $s_u$ sends a "turning-on" signal to unblock site $s_p$, allowing spider $Y = 1$ to take the unblocked central path from site $s_p$ to the output location. Under other input assignments, the 1-valued spider gets stuck at the crossroad, so only the 0-valued spider can reach the output location. Therefore, the AND gate yields 1 when both spiders are assigned 1, and yields 0 in all other cases. Similarly, in the OR gate, when both spiders are 0, they meet at the crossroad in the middle and only spider $Y = 0$ can reach the output location. Under other input assignments, the 0-valued spider gets stuck at the crossroad, so only the 1-valued spider can reach the output location. Therefore, the OR gate yields 0 when both spiders are assigned 0, and yields 1 in all other cases.

route is cut off. We explain how mechanism $SW_{1 \to 0}$ works with a 1-valued spider as an example; mechanism $SW_{0 \to 1}$ works analogously.

Mechanism $SW_{1 \to 0}$ is formed by three neighboring functional sites along the horizontal direction: $s_{1 \to 0}, s_r^I, s_r^{II}$. We use a staging transition diagram in Fig. 5 to describe how mechanism $SW_{1 \to 0}$ changes a 1-valued spider to be 0, and cuts off the backward route of the spider. A stage transition shows the change of the spider's location, value or the site states. At stage (1), all sites are "on" initially. Site $s_{1 \to 0}$ can trap a spider, and contains a "switching to 0" signal that will be sent to its left site when a spider attaches to it. Therefore, when a 1-valued spider attaches to $s_{1 \to 0}$, it is trapped and receives the signal changing its value to 0, causing a transition to stage (2). At stage (2), since the limb trapped at site $s_{1 \to 0}$ cannot move back, the spider could only move forward by attaching to site $s_r^I$ that traps the spider and sends out a "turning off" signal to its left site. When site $s_{1 \to 0}$ receives that signal and turns itself "off", we get to stage (3). At stage (3), the limb trapped on $s_r^I$ cannot move back, the spider could only

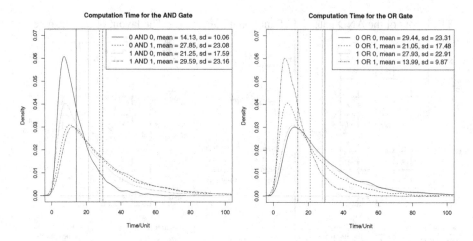

**Fig. 4.** The computation time distributions for the AND gate and the OR gate under four possible input assignments. Each curve in one gate represents a time distribution under one assignment. The vertical line indicates the mean value of computation time under one assignment in the simulation. The standard deviation for each curve is shown in the legend.

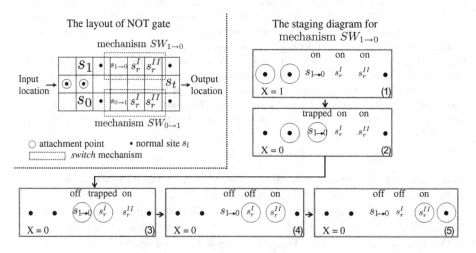

**Fig. 5.** The layout of gate NOT is shown in the figure. The function of mechanism $SW_{1\rightarrow0}$ is to switch a spider's value from 1 to 0 and cuts off its backward route. We show how mechanism $SW_{1\rightarrow0}$ works in a staging transition diagram, where the spider value is expressed as $X$ and the state of each functional site is shown above it.

move forward by attaching to site $s_r^{II}$ that sends a "turning off" signal to its left site. When $s_r^I$ receives that signal and turns itself "off", we get to stage (4). At stage (4), the limb on $s_r^I$ can transit to a normal site on the right of $s_r^{II}$, while the limb on $s_r^{II}$ cannot move back to $s_{1\rightarrow0}$ which is "off". The spider could only

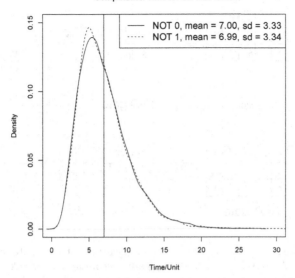

**Fig. 6.** The computation time distributions for the NOT gate under two possible input assignments.

move forward to get to stage (5). At stage (5), sites $s_r^I$ and $s_r^{II}$ are "off", the spider cannot walk back. When a spider goes through these 5 stages, its value is switched and its backward route is cut off. The mechanism $SW_{0\to1}$ comprising $s_{0\to1}, s_r^I, s_r^{II}$ follows similar staging transitions, the only difference being that a 0-valued spider becomes 1 in the stage transition (1) to (2).

Figure 6 shows the computation time distributions for the NOT gate. The distribution curves for the two input assignments are long-tailed and alike, which is due to the symmetric path design for the 1-valued spider and the 0-valued spider.

## 3.4   Gate Cascades

To construct a large logic circuit, we need to cascade logic gates of the three kinds defined in Sects. 3.2 and 3.3. A wire $w$ connecting an upstream gate and a downstream gate is composed of continuous normal sites $s_l$. To ensure that the spider that reaches the output location exits the upstream gate and never goes back to it, we place two additional sites $s_r^I$ and $s_r^{II}$ after site $s_t$ on the output location, forming an *exit* mechanism which cuts off the backward route of a spider that moves through it.

The mechanism *exit* follows similar staging transitions to mechanism $SW_{1\to0}$ shown in Fig. 5. It consists of three neighboring functional sites along the horizontal direction: $s_t, s_r^I, s_r^{II}$. We explained the functionality of site $s_r^I$ and $s_r^{II}$ at the end of Sect. 3.3. Site $s_t$ is designed to trap the spider. Therefore, a staging transition diagram for mechanism *exit* is similar to the one shown in Fig. 5,

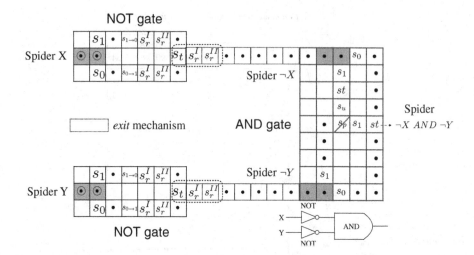

**Fig. 7.** A logic circuit: $(\neg X \wedge \neg Y)$. The input locations of each gate are highlighted in grey. Spiders $X$ and $Y$ exit the NOT gate, becoming spider $\neg X$ and $\neg Y$ after passing through the *exit* mechanisms. The AND gate computation begins whenever a spider enters the AND gate. The spider reaching the output location of the AND gate represents the computation result $\neg X \wedge \neg Y$.

with the only difference that the spider value is unchanged throughout the five stages. For a downstream gate with two inputs, its two input spiders may arrive at different moments. Computation of the downstream gate begins when either input spider enters the gate, and the asynchronous arrival of input spiders will not influence the computation accuracy of the gate.

Figure 7 illustrates a simple logic circuit implemented by cascading two NOT gates as the inputs to an AND gate. The output location of each NOT gate is connected to an input location of the AND gate via the *exit* mechanism. Spider $X$ and spider $Y$ start to move in the two NOT gates concurrently. When the two spiders move out of the NOT gate, their backward routes are cut off due to the *exit* mechanisms, and they have their values changed to $\neg X$ and $\neg Y$. When either spider enters the AND gate, gate computation begins, yielding the result $\neg X \wedge \neg Y$ eventually. The computation time of this logic circuit is shown in Fig. 8. In all simulation runs, the output spider produced the correct output value.

### 3.5   Complexity Analysis

In a single gate, the computation time $t_{\text{gate}}$ is the traversal time of the spider that reaches the output location. Since the spider moves on the track stochastically, the computation time $t_{\text{gate}}$ is a random variable following a long-tailed distribution, as shown in Figs. 4 and 6.

When a spider leaves a gate or enters a gate, its backward route is cut off due to the functionality of the *exit* mechanism, so we can use the computation

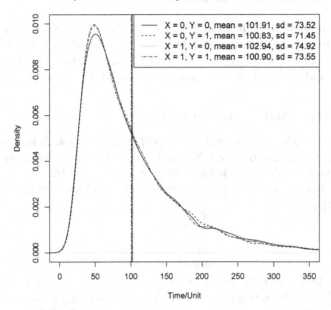

**Fig. 8.** The computation time distribution for the logic circuit $(\neg X \wedge \neg Y)$ under four possible input assignments.

time of a single gate $t_{gate}$ to estimate the computation time $t$ of a circuit. For any $n$-variable boolean function, we can transform it into 3-CNF, which is a conjunction of $m$ clauses, each a disjunction of at most three literals. Since our design allows parallel evaluation, for a clause $m_i = (l_1^i \vee l_2^i \vee l_3^i)$, the computation time of $m_i$ is

$$t_{m_i} \leq 2 \times (t_{\text{OR}} + t_{\text{NOT}}) = O(1).$$

Since each clause needs time $t_{m_i}$, to evaluate $m$ clauses in parallel, we conduct $\log m$ AND gate computations that cost $t_{\text{AND}} \times \log m$, and in total use time

$$t = t_{\text{AND}} \times \log m + t_{m_i} = O(\log m).$$

For any boolean function in 3-CNF with $m$ clauses, we use at most $3m$ spiders to represent the literals. For each clause, we need at most three NOT gates and two OR gates if all the literals are the negation of a variable, which is a constant number. For $m$ clauses, we need $m - 1$ AND gates. Therefore, the total space complexity is $O(m)$. Hence, our circuit designs are scalable because circuit size in our design scales linearly with formula size, and evaluation time is logarithmic in the formula size.

# 4    Formal Definition of the Model

The active molecular spider system is modeled as a continuous-time Markov process where the state transitions depend on the interactions between the molecular spiders and the sites on the track. We first define the site types and transition rules of alterable sites, and then give a formal definition of the model.

## 4.1    Site Types and Transition Rules

Sites are categorized into *normal sites* and *functional sites*. A normal site $s \in S_{norm} = \{s_l, s_0, s_1\}$ has no state. Site $s_l$ binds to the spider's leg. Sites $s_0$ and $s_1$ bind to the spider's arm if it has type 0 or 1, respectively.

A functional site $s \in S_{fun}$ has a state of "on", "off" and "trapped". The site state transition diagram is:

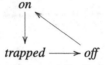

A spider limb can only attach to an "on"-state site. An "off"-state site is non-alterable. The limb trapped on a "trapped"-state site cannot leave the site by itself. Whether a site can trap a spider is indicated by $TR \in \{0,1\}$: a site with $TR = 1$ will trap a spider when a limb attaches to it. A functional site may change the spider's value, or the state of another site, by sending out a *signal* to the spider or another site. We define

$$signal = (val, d) \text{ or } null, \text{ where } d \in \mathbb{Z}^2 \text{ and } val \in \{on, off, trapped, 1, 0\}. \quad (1)$$

Suppose a functional site is located at $(x, y)$. If it holds a $signal = (val, d = (d_x, d_y))$ then it sends the signal to the location $(x + d_x, y + d_y)$, setting the state of the site located there, or the spider's value, to $val$. When $d = (0,0)$, the $val$ field of the signal is either 1 or 0, which is sent to the spider, setting the spider's value to 1 or 0.

Therefore, we can define a functional site $s \in S_{fun}$ as

$$s = (state, TR, signal). \quad (2)$$

The signal held in a site is sent out once a spider limb attaches to the site. When a signal is sent out, the site has no signal remaining, which we express as $s = (state, TR, null)$. A functional site $s = (on, null)$ is equivalent to a normal site, which is non-alterable. Once a signal is received by a site or a spider, the site state or the spider's value is changed according to the signal.

In the logic circuit construction, we use two functional sites $s_u$ and $s_p$ in the AND gate and OR gate, and we design a set of functional sites that form different mechanisms in the NOT gate and the gate cascades. Table 1 gives the definitions of these functional sites and the *transition rules* applied to them. A functional site $s$ transits to site $s'$ in the second column, either by receiving a signal or being attached by a spider limb. If $s$ holds a signal, it causes other changes in

**Table 1.** Definition of different functional sites used in the circuit construction and the *transition rules* applied to them. Suppose the location of the site is $(x, y)$, define $(x', y') = (x + d_x, y + d_y)$.

| Transition rules | | |
|---|---|---|
| Functional site | Updated site | Other changes |
| $s_t = (\text{on}, 1, null)$ | $s'_t = (\text{trapped}, 1, null)$ | |
| $s_{1 \to 0} = (\text{on}, 1, (0, (0, 0)))$ | $s'_{1 \to 0} = (\text{trapped}, 1, null)$ | $A = 0$ |
| $s_{0 \to 1} = (\text{on}, 1, (1, (0, 0)))$ | $s'_{0 \to 1} = (\text{trapped}, 1, null)$ | $A = 1$ |
| $s_r^I = (\text{on}, 1, (\text{off}, d))$ | $s_r^{I'} = (\text{trapped}, 1, null)$ | site at $(x', y')$ becomes off |
| $s_r^{II} = (\text{on}, 0, (\text{off}, d))$ | $s_r^{II'} = (\text{on}, 0, null) = s_l$ | site at $(x', y')$ becomes off |
| $s_u = (\text{on}, 0, (\text{on}, d))$ | $s'_u = (\text{on}, 0, null) = s_l$ | site at $(x', y')$ becomes on |
| $s_p = (\text{off}, 0, null)$ | $s'_p = (\text{on}, 0, null) = s_l$ when a "turning-on" signal is received | |

the last column. In Table 1, the updated site $s'$ in the second column is either a normal site or a trapped site. According to the site state transition diagram, a trapped site can only transit to a "off"-state site that is non-alterable by itself. Since no signals are designed to turn on these "off"-state sites transited from the trapped sites, these "off"-state sites are non-alterable finally. Therefore, all the functional sites in Table 1 are alterable initially and become non-alterable finally. The functional sites used in our design are

$$\{s_t, s_{1 \to 0}, s_{0 \to 1}, s_r^I, s_r^{II}, s_u, s_p\},$$

where each site $s$ among them includes its site transitions under the *transition rules* described in Table 1. The set of site types is $S = S_{norm} \cup S_{fun}$.

A *mechanism* is a set of neighboring mechanism sites along the same direction. We design three different mechanisms used in the logic circuit construction. The *switch* mechanism $SW_{1 \to 0}$ ($SW_0 \to 1$) contains sites $s_{1 \to 0}(s_{0 \to 1}), s_r^I, s_r^{II}$, where site $s_r^I, s_r^{II}$ contains the signal of $(\text{off}, (-1, 0))$ which can block its left site. When a spider moves over the *switch* mechanism, its value is flipped, and its backward route is cut off. The *exit* mechanism contains sites $s_t, s_r^I, s_r^{II}$. When a spider moves over this mechanism, its backward route is cut off.

When a spider limb leaves a site, this limb can reach 4 sites geometrically (shown in Fig. 1). Since sites have different types, wether a site is available for a limb of a spider depends on the spider value and the site types. Given a spider with value $A$ and a site, algorithm *check* summarizes how to tell if the site is available.

Using Algorithm 1, we examine every site among the 4 sites shown in Fig. 1, putting those available into a set $\mathscr{AV}$.

## 4.2 Model Definition

The *active* multi-spider system with normal sites and alterable sites can be modeled as a continuous-time Markov process. We define the state of the model as

$$X = (S_1, S_2, \ldots, S_n, E), \tag{3}$$

---

**Algorithm 1.** Algorithm *check* tells if a given site is available.

---

- if the site is occupied by another spider, it is not available.
- else:
  1. if the site is a normal site:
     (a) if the site is $s_l$, it is available;
     (b) if the site is $s_1$ and $A = 1$, it is available;
     (c) if the site is $s_0$ and $A = 0$, it is available;
  2. else if the site is a functional site:
     (a) if the site is $s_{1\to0}$ and $A = 1$, it is available;
     (b) if the site is $s_{0\to1}$ and $A = 0$, it is available;
     (c) if the site is "on"-state, it is available;
  3. else, the site is not available.

---

where $S_i = (P_i, A_i)$ $(1 \leq i \leq n)$ describes the state of the $i$-th spider. Set $P_i = (p_a^i, p_b^i)$ contains attachment points for the $i$-th spider, and $A_i \in \{0, 1\}$ represents the Boolean value of the spider. The lattice configuration $E : \mathbb{Z}^2 \to S$ shows the layout of different sites, where $S$ is the set of site types. Normal sites can be regarded as having state "on", $TR = 0$ and no signal, so we can redefine the lattice configuration as

$$E : \mathbb{Z}^2 \to \{\text{on, off, trapped}\} \times \{1, 0\} \times \mathbb{S},$$

where $\mathbb{S}$ represents the set of signals.

Given a model state $X = (S_1, S_2, \ldots, S_n, E)$ at time $t$, if a limb leaves an attachment point $p \in P_i \in S_i$, we use the algorithm *check* to obtain a set of available sites $\mathscr{AV}$. At time $t + \delta$, this limb transits to $p' \in \mathscr{AV}$, changing the set of attachment points to $P_i' = P_i - \{p\} \cup \{p'\}$. We use the *transition rules* to update $A_i$, so we have $S_i' = (P_i', A_i')$. The transition rules also updates $E$, thus the new state is

$$X' = (S_1, S_2, \ldots, S_{i-1}, S_i', S_{i+1}, \ldots, S_n, E').$$

## 5   Conclusions and Discussions

Using an *active* multi-spider model with spider cooperation and localized signal transmission, we have implemented the basic logic gates (AND, OR, NOT). We have shown how to implement gate cascades, in which each upstream gate $G_u$ is connected to a downstream gate $G_d$ using the *exit* mechanism. We use $O(1)$ types of spiders and sites. To evaluate an $n$-variable Boolean function that is in 3-CNF with $m$ clauses, the evaluation time is $O(\log m)$ and the size of the circuit is $O(m)$. Therefore, our design supports scalable computation and ensures spatial locality. Molecular circuits with spatial locality overcome the challenges of computation speed-up and sequence reuse in molecular computing in a well-mixed environment, but there are still other issues. Compared with previous work [2,3,5], our design better addresses the following issues:

**Geometrical Layout.** Molecular circuits with spatial locality arrange different computing components on a 2D plane where the distance between different components should be set carefully to avoid interference across components. Reducing the number of gates used in a circuit can ease the geometrical layout problem. Our design implements a NOT gate to avoid dual-rail logic conversion used in previous work [2,5], which simplifies the circuit and its layout. Compared with the circuit [3] in a form of BDD where the layout of different branching paths requires appropriate angles and lengths, our design only considers connections between gates because each gate has a fixed layout.

**Data Encoding.** In previous work, variable representation is encoded into the circuit [2,3,5], so each variable corresponds to a distinct sequence. This complicates sequence design if the circuit has a large number of variables. Our design separates variable representation from circuit design, only using two types of spiders placed at different input locations to represent all variables.

**Circuit Reusability.** Tethered circuits [2,5] use irreversible local signal transmission to implement logic computation and value propagation, so the circuit is not reusable. The circuit [3] adds external strands to unblock a path for an evaluating walker. This procedure irreversibly changes the circuit configuration, thus the circuit is not reusable. In our design, irreversible local signal transmission is used to control the spiders' behavior at a few locations in the circuit, which only occupy a small portion of computation. Since non-alterable sites form the majority of the circuit, most parts of the circuit are reusable.

We lack an experimental implementation of our designs, thus here we use a simulator that simulates the circuit at the site level, assuming spiders have equal transition rates to all sites. We are working on an implementation where normal sites are short DNA strands so that molecular spiders can attach to or detach from the normal sites freely, and functional sites transmit signals to neighboring sites via strand displacement. For example, we can encode a signal in the loop (inactive part) of a hairpin structure. Once a spider attaches to the hairpin structure, the loop is opened so that the exposed domain can react with other neighboring sites, transmitting the signal encoded in the opened loop to other neighboring sites. In the future, we will complete a plausible implementation and focus on a simulator that can better reflect how different sites react with spiders according to that implementation.

Since spiders move bidirectionally on the track, we can use this feature to solve some interesting problems. For example, it may be possible to construct a feedback loop that can be used to solve a SAT problem automatically where the spider that does not satisfy the formula can go back to switch its value. In the current model, molecular spiders can probe, walk, and change their own states and the state of the environment. These behaviors of the molecular spiders can be extended for complex intracellular tasks, e.g., we can use the molecular spiders to replace natural motors. In the future, we will explore applications of our design, as well as the possibility of implementing it in the laboratory.

# References

1. Antal, T., Krapivsky, P.L.: Molecular spiders with memory. Phys. Rev. E **76**(2), 021121 (2007)
2. Chandran, H., Gopalkrishnan, N., Phillips, A., Reif, J.: Localized hybridization circuits. In: Cardelli, L., Shih, W. (eds.) DNA 17 2011. LNCS, vol. 6937, pp. 64–83. Springer, Heidelberg (2011)
3. Dannenberg, F., Kwiatkowska, M., Thachuk, C., Turberfield, A.J.: DNA Walker circuits: computational potential, design, and verification. In: Soloveichik, D., Yurke, B. (eds.) DNA 2013. LNCS, vol. 8141, pp. 31–45. Springer, Heidelberg (2013)
4. Lund, K., Manzo, A.J., Dabby, N., Michelotti, N., Johnson-Buck, A., Nangreave, J., Taylor, S., Pei, R., Stojanovic, M.N., Walter, N.G., et al.: Molecular robots guided by prescriptive landscapes. Nature **465**(7295), 206–210 (2010)
5. Muscat, R.A., Strauss, K., Ceze, L., Seelig, G.: DNA-based molecular architecture with spatially localized components. ACM SIGARCH Comput. Architect. News **41**, 177–188 (2013). ACM
6. Padilla, J.E., Liu, W., Seeman, N.C.: Hierarchical self assembly of patterns from the Robinson tilings: DNA tile design in an enhanced tile assembly model. Nat. Comput. **11**(2), 323–338 (2012)
7. Pei, R., Taylor, S.K., Stefanovic, D., Rudchenko, S., Mitchell, T.E., Stojanovic, M.N.: Behavior of polycatalytic assemblies in a substrate-displaying matrix. J. Am. Chem. Soc. **128**(39), 12693–12699 (2006)
8. Qian, L., Winfree, E.: Scaling up digital circuit computation with DNA strand displacement cascades. Science **332**(6034), 1196–1201 (2011)
9. Samii, L., Blab, G.A., Bromley, E.H.C., Linke, H., Curmi, P.M.G., Zuckermann, M.J., Forde, N.R.: Time-dependent motor properties of multipedal molecular spiders. Phys. Rev. E **84**, 031111 (2011)
10. Semenov, O., Mohr, D., Stefanovic, D.: First-passage properties of molecular spiders. Phys. Rev. E **88**(1), 012724 (2013)
11. Semenov, O., Olah, M.J., Stefanovic, D.: Mechanism of diffusive transport in molecular spider models. Phys. Rev. E **83**(2), 021117 (2011)
12. Semenov, O., Olah, M.J., Stefanovic, D.: Multiple molecular spiders with a single localized source—the one-dimensional case. In: Cardelli, L., Shih, W. (eds.) DNA 17 2011. LNCS, vol. 6937, pp. 204–216. Springer, Heidelberg (2011)
13. Wickham, S.F., Bath, J., Katsuda, Y., Endo, M., Hidaka, K., Sugiyama, H., Turberfield, A.J.: A DNA-based molecular motor that can navigate a network of tracks. Nat. Nanotechnol. **7**(3), 169–173 (2012)

# Organic Mathematics: On the Extension of Logics from Physical Atoms to Cellular Information Processes

Jerry L.R. Chandler[(⊠)]

Krasnow Institute for Advanced Study, George Mason University,
Fairfax, VA, USA
Jerry_LR_Chandler@Mac.com

**Abstract.** Formally, cellular information processing requires mathematical forms of representation of parts of the whole. Several types of mereological operands, co-operands, and operators are necessary to embed the symbolic meanings. The common source of propositions is the atomic numbers, taken as both (measurable) physical objects and mathematical objects. The atomic numbers serve as source of the novel notation for organic mathematics. The historical scientific basis for the notation is described in terms of the electrical particles (abstractly defined as cardinal and ordinal numbers). The regular order of atomic numbers as diagrams define the radix attributes of the perplex number system. A triad of mathematical constructs built from the perplex numerals include quantum mechanics, the table of elements, and molecular biology. The radix of the hybrid logic of biological information processing includes forms of both copulative and predicative propositions on the attractive and repulsive attributes of electrical particles.

**Keywords:** Organic mathematics · Perplex numbers · Chemical notation · Synductive logic · Mereology · Constructive mathematics · Electrical relations · Categories

## 1 Introduction

### 1.1 The Symbolic Gap

A quantitative theory of biological information processing requires a general method to relate the biological changes of organic structures to orthodox mathematical and physical symbols. But, the notation for chemistry and that of mathematics use two different numbering systems, two different modes of logical extension. The latter bases extension on Pythagorean geometry. The former number system extends the count by generating new relation classes from the natural reference system. This "difference that makes a difference" between the chemical and mathematical symbol systems create an insurmountable logical gap between the disciplines. The radix notation for biological information processing emerges from the algebra of chemical symbols in contrast with the informational processing of physical systems. The communications gaps arise as a consequence of the incompleteness of the logical expressive capacity of orthodox

© Springer International Publishing Switzerland 2015
M. Lones et al. (Eds.): IPCAT 2015, LNCS 9303, pp. 29–35, 2015.
DOI: 10.1007/978-3-319-23108-2_3

mathematics with respect the description of chemical and biological structures and processes (see Chandler et al. 1995).

## 1.2   The Logical Gap

The logic of chemical algebra differs from the logic of physical algebra. Organic Mathematics is the formal mathematics of the chemical notational system (Chandler 2014). Organic objects are expressively complete with respect to the atomic numbers. Informally, the logic of organic mathematics exists now, encoded within the copulative grammar of chemical and molecular biological propositions and the logical operations on molecular formula, molecular structures, metabolites, biomolecule synthesis, inheritance, adaptation, and so forth. The radix of cellular information processing is necessarily expressed in the specific electrical relations describing the existence of the organic components of the organization of the organism. A critical distinction between physical mathematics and organic mathematics is the crisp separation of the foundational chemical concept of matter from the foundational scientific concepts of space and time. The novel electrical relations of organic mathematics bridge many of the gaps of meaning that separate the mathematical sciences from chemical sciences. In the recent book "Integral Biomathics" (Simeonov et al. 2012), several prominent mathematicians and informational scientists called for a new mathematics for biology. Although they did not provide specific criteria for identifying "biomathics", the foundational logic of Organic Mathematics provides a coherent testable hypothesis that can be evaluated against future criteria, should they become available.

## 2   A Novel Electrical Notation for the Bio-chemical Sciences

### 2.1   Origins of Atomic Numbers

Organic mathematics introduces a novel notation for the chemical sciences that is expressively complete and deductive with respect to both mathematical structures and physical laws. The novel notation is necessary for a formal basis for a scientifically coherent expression of such vital cellular information processes as metabolism, inheritance, and emergence. The logical construction of organic mathematics is initiated by the application of Newton's and Coulomb's laws to individual chemical elements, resulting in the induction of the term "atomic number" (Born 1969) to represent the electrical relations within every atom. Hence, the radix of organic mathematics is a propositional term logic referenced in the atomic numbers as constructs of electrical particles. Three different mathematical constructions of mereological relations have emerged from the physical conceptualization of these natural integers. The entelechy of each of the mathematical constructs is to describe part-whole relations in order correspond with and validate empirical measurements on natural systems.

## 2.2    Extensions of the Atomic Numbers

The mereology of atoms as electrical dynamical systems, the mereology of molecules as physical compositions of multi-sets of atomic numbers, and the mereology of cells with physical potential to reproduce the multi-set of compounds inherited from progenitors are all based on the abstract concept of the linear order of the atomic numbers. Roughly speaking these three types of part-whole relations infer the intrinsic propositional terms of "quantum mechanics", "chemical synthesis" and "molecular genetics". The three mathematical compositions are guided and constrained by the common physical principles of the conservation of matter and the conservation of electricity. These two physical conservation laws necessitate exact indices for each unique whole that contains every individual part of a whole. Logically and necessarily, each part of a whole enters into the mathematical construct, otherwise the mathematical construct would fail Hilbert's criteria of consistency, completeness and decidability for mathematics, and would fail the physical conservation laws and would fail, roughly speaking, to reproduce the multi-set of components of the progenitors. In other words, in order to ensure scientific coherence among parts of wholes, the unit of a whole must be a logical consequence of the units of the parts. The electrical units of the parts of an atom (positive nuclei, negative electrons) entail the mereology of atomic numbers (Quantum mechanics describes the motion of parts of atoms in relation to energy units.). The units of the parts of a molecule are the atomic numbers, taken together as the molecular formula representing all the electrical parts of the whole (An identity theorem is necessary for each molecule.). The unit of a cell infers a specific object that can be expressed within a specific multi-set of components such that reproduction is entailed through dynamic organic feedback and feed forward relations (The internal mereological relations of a cell are a consequence of both the antecedent physical environment and its antecedent organic structures. The on-going information processing of a living cell is thus interdependent on the relations between both internal and external organic structures.). In summary, despite the differences in mathematical forms, mathematical functions and logical operators, **the triad of mereologic constructs from the atomic numbers are logically consistent with one another.** The notation for organic mathematics captures the consistency among part-whole relations among relatives.

# 3    Perplex Notation and Physical Dependence of Organic Operands and Operations

## 3.1    Physically Independent and Interdependent Electrical Particles

The proposed notation for organic mathematics represents the physically *independent* atomic numbers as a set of relatives (e.g., the chemical table of elements), as members of a closed multi-set (Chandler 2009). A molecular number is represented as an *interdependent* multi-set of relatives. The logical operation that transforms independent objects (operands) to interdependent objects forms measurable physical links between the relatives by emergent electrical relations. Hence, the logical expressive power of organic mathematics is restricted to electrical relations between electrons and nuclei.

Abstractly, the organic operators may conjoin any two integers. Pragmatically, physical evidence is necessary to demonstrate the attributes of each emergent class.

This novel notation for both mathematics and chemistry was introduced as the perplex number system (Chandler 2009). The meaning of the notational symbols reflects the various roots of the mathematical, physical and chemical concepts from which the number system is constructed (Chandler 2009). The heterodoxy of organic mathematics was contrasted with orthodox mathematics (of category theory) in algebraic biology (Chandler 2009).

### 3.2   Constraints on Formal Electrical Logics

The formal propositional logic of organic mathematics is excluded from all standard forms of logic that I am aware of *because* the meaning of the terms includes the polarity of electrical relations among relatives. Classical chemical notation developed as an expression of physical data as chemical diagrams of relationships among relatives. It necessarily requires an identity term to specify each unique physical entity of the chemical plexus. This identity term is synonymous with a unique rhetorical name as well as the components of the composition (Chandler 2014). Furthermore, the perplex logic of organic mathematics must preserve the organic indexing of the empirical associations of meanings of number symbols such that organic diagrams express the same meaning as the orthodox chemical notation. *Several indices may be requires to express the relative places in relative spaces, for example, to express the multiple handednesses of bio-macromolecules.*

The meaning of a perplex numeral symbol differs from the meaning of a simple number symbol as well as that of the notation for a complex number. The orthodox meaning of an integer number symbol may correspond with the geometric distances (Pythagorean Theorem) or Peano's postulates. The meaning of a complex number is defined in terms of the square root of a minus one ($-1$). *The meaning of a perplex numeral is defined as a triad of concepts, an ordinal number, a cardinal numbers and an equinumerous set of relations to represent an atomic number.* This is logically the same as a star graph with a central ordinal connecting a corresponding set of cardinals. This is a physically motivated definition, inferring a correspondence relation between mathematical and physical concepts. Physically, this definition corresponds with counts the electrical charges of the mereological parts of an atom. These physical constraints are interpreted as mathematical constraints on the logic of ordinal and cardinal numbers as operands and operators.

## 4   Mereology and Electrically Labeled Bipartite Graphs

### 4.1   Ratiocinations

A central postulate of the organic notation is that the electrical attributes of matter enumerate the structures of organic mathematics, its operands and its operators, analogous to the role of electrical properties of each chemical element for enumerating its internal structure. This is a critical assertion for the organic sciences for three specific ratiocinations. Prior to the 20th Century enumeration of the electrical parts of an atom, Lavoisier (1778) and Dalton (1806) grounded chemical notations in the physical attribute of weight (or mass). By switching the radix (base) of the notation to the electrical regularity of the atomic numbers as perplex numerals and numbers, **the novel notation allows a simple algebra to be developed for compounding integers and ratio of integers into irregular patterns.** Such irregular patterns can be enumerated in logical terms as perplex numbers, perplex diagrams (icons, structures) and an irregular tabular arrangement, patroids (roughly homologous to matroids). Secondly, the ratiocinations from this critical assertion constrains the associations of terms of perplex logic. Electrical attributes of the parts of atoms are either mutually attractive or mutually repulsive. Pairs of perplex ordinal numbers are mutually repulsive. Pairs of cardinal numbers are mutually repulsive. Only a pair of an ordinal number with a cardinal number forms an attractive unit and hence can be logically associated adjacent to one another. These facts of pairings of electrical particles as attractors and repellors are crucial to physical quantum theory as well. *(The reader should note the difference between electrical pairings of ordinal and cardinals and set theory pairings and the logical implications thereof.)* Thirdly, this assertion establishes the logical foundation for emergence and submergence of relations among relatives, for the economy of radix relations essential to the representation of biological reproduction in logical terms.

## 4.2   Mereology

The first level of logic of organic mathematics places independent but relative objects in relationships to one another *as parts of a whole.* The chemist's "arrow" (as an operator) signifies the placements are to be related as nodes of graphs in a three dimensional network. Each perplex numeral (atomic number) *represents a star graph in the iconic notation of organic mathematics.* Each unit of the chemical plexus is a composition of star graphs, as signified *by a labeled bipartite graph that represents its identity.* The construction of a whole from its parts (composition of molecular numbers from atomic numbers) is guided by valence and other local physical situations. Under the conservation laws of physics, the bijective mapping from a multi-set of perplex numerals to a type of the chemical plexus preserve the parts of the whole. Attraction and repulsion among the ordinal and cardinal components of a multi-set "glue" the parts into a harmonious whole by generating new inter-graphic edges. *The number and the placement of newly generated edges form a fresh identity by connecting the independent organic objects into a singular interdependent object.* The emergent interdependent object is synonymous with an organic structure, a logical diagram and a patroid, a tabular representation of the data that generates the chemical icon.

## 5   Conclusions

### 5.1   The Role of Hybridizations of Logics in Organic Mathematics

The novel "type" notation of organic mathematics forms a hybrid associative logic for the construction of logical diagrams from fragments (see: Brauner 2011). Roughly speaking the perplex notation hybridizes propositional sentences originating from mathematical, physical and chemical theories. The essential meanings of the arithmetic logic of ordinal and cardinal numbers, the predicate logic of physical attributes, and the copulative logic of organic proofs of structure is captured by this type notation. This form of local associative extension, guided by the order of numerals, the individual electro-neutrality values of each integer and its potential valences are extendable indefinitely to any integer. *Emergence of perplex objects (molecules, bio-macro-molecules, genetic systems...) is quantified within diagrams of the chemical plexus.* The major scientific advantage of the novel notation of organic mathematics over other approaches to theories of complex systems lies in the desmology. The quantities of ligands can be extended over the integers by copulative logic based on the universal role of the atomic numbers and validated physical measurements such as x-ray crystallography. The notation can provide the initial conditions for theories of dynamical systems, including quantum mechanics. (Szabo and Ostlund 1996, pp. 40–46).

### 5.2   Future Work and International Collaboration

This work originated in 1972 from attempts to describe the relationships between chemical structures and cellular information processing. We were studying the triadic relationships between chemical reaction mechanisms of alkylating agents, DNA reaction sites and the forms of mutagenic does-response relationships. The problem remains open today. This work emerges from these initial efforts to relate structure-function relations among chemical relatives. This note gathers together some of logical aspects of the relevant scientific theories into a coherent notation for the identities of individuals. The notation provides a radix (base) for the ampliative logic for construction of new organic terms.

The sortal logic of types can be viewed as the spine of the economies of relations among organic relatives. Such sortal logics of types, by preserving the ordering principle of the linearity of the numerals in the logical size of objects (mereological part/whole compositions), can be extended to any integers by addition. From the radix of perplex numerals, new mathematical objects (graphs, forests of graphs, etc.) can be extended indefinitely to higher order organizations of matter by sublations on collections of terms. This fact opens opportunities for explorations of outstanding problems of mathematical biology.

I seek international colleagues from the logical, mathematical, physical, biomedical and computer sciences to collaborate on these challenging problems.

## 6   Summary

In summary, the logic intrinsic to the atomic numbers can now be used to construct a unique diagrammatic/iconic logic and organic terms that can contribute to bridging the communications gaps between mathematics and biological sciences.

# References

Born, M.: Atomic Physics, 8th edn., p. 63, 127. Dover, New York (1969)

Brauner, T.: Hybrid Logic and Its Proof Theory. Springer, Berlin (2011)

Chandler, J.L.: Introduction to the Perplex Number System. Discrete Appl. Math. **157**, 2296–2309 (2009a)

Chandler, J.L.R.: Algebraic biology. Axiomathes **19**(3), 297–320 (2009b)

Chandler, J.L.R.: A systematic proposal for a philosophy of natural identities. In: Proceedigns of 12th Special Focus Symposium on Art and Science. IIAS, pp. 57–60 (2014)

Chandler, J.L.R., Ehresmann, A.C., Vanbremeersch, J.P.: Contrasting two representations of emergence of cellular dynamics. In: Symposium on Emergence (1995). http://cogprints.org/1659/1/Baden%2DBaden_95.htm

Igamberdiev, A.U.: Physics and the Logic of Life, pp. 103–127. Nova Science, Hauppauge (2012)

Karplus, M., Porter, R.N.: Atoms and Molecules. Benjamin, London (1970)

Simeonov, P.L., Smith, L.S., Ehresmann, A.C.: Integral Biomathics. Springer, Berlin (2012)

Szabo, A., Ostlund, N.: Modern Quantum Chemistry, pp. 40–46. Dover, New York (1996)

# Collective and Distributed Behaviour

# An Ecosystem Algorithm for the Dynamic Redistribution of Bicycles in London

Manal T. Adham$^{(\boxtimes)}$ and Peter J. Bentley$^{(\boxtimes)}$

University College London, London, UK
m.adham@cs.ucl.ac.uk, peter@peterjbentley.com

**Abstract.** We extend and adapt the Artificial Ecosystem Algorithm (AEA), by applying it to the dynamic redistribution of bicycles in London's Santander Cycle scheme. Just as an ecosystem comprises many separate components that adapt to form a single synergistic whole, the AEA uses a bottom up approach to build a solution. A problem is decomposed into relative subcomponents, they then evolve and cooperate to form solution building blocks, which connect to form a single optimal solution. In this way the AEA is designed to take advantage of highly distributed computer architectures and adapt to changing problems. Three variants of the AEA are described and applied to the Santander Cycle scheme: AEA, AEA Random and AEA Nearest Neighbour. The algorithms have been tested using historical data and empirical results prove their potential effectiveness.

## 1 Introduction

London's Santander Cycle scheme currently operates 748 docking stations and 11500 bikes [13]. Forming a large component of London's transport system. Due to usage patterns, there is often an imbalance between user demand and the service provided by the scheme [17]. To overcome this issue and ensure constant availability of bikes/spaces at docking stations, a fleet of trucks manually redistribute bicycles.

As the complete input for the dynamic redistribution of bikes is not known in advance, exact methods cannot be used, in retrospect we use biology inspired algorithms to find practical optimal solutions. Dynamic vehicle routing is a similar problem [2], which has been addressed in multiple papers [5,7,8,16]. However, the algorithms proposed typically deal with small network sizes and it is difficult to scale them to hundreds of docking stations and thousands of bicycles.

In this work we apply the Artificial Ecosystem Algorithm (AEA), which was used to solve the travelling salesman problem (TSP) in [1]. The process of determining routes for a fleet of vehicles allows you to select any sequence of docking stations, therefore the problem grows exponentially with the number of docking stations, making this a combinatorial optimisation problem similar to the TSP [4].

The AEA accepts that some classes of problems can become so prohibitively large, that it is more appropriate to divide the problem into smaller more

© Springer International Publishing Switzerland 2015
M. Lones et al. (Eds.): IPCAT 2015, LNCS 9303, pp. 39–51, 2015.
DOI: 10.1007/978-3-319-23108-2_4

tractable pieces and have separate processors work on them. Most evolutionary algorithms such as Genetic Algorithms [6] represent an entire candidate solution in each individual of the population, any attempts to distribute the algorithm across many processors will divide populations, not individual solution evaluations.

AEA solves a problem by adapting its subcomponents such that they connect to form a single optimal solution, akin to the way an ecosystem comprises many separate components that adapt to form a single synergistic whole. Like the different species in an ecosystem, the AEA may have species of components representing sub-parts of the solution that evolve together and cooperate with each other. In this way the AEA is designed to take advantage of highly distributed computer architectures and adapt to dynamic problems.

This paper is organised as follows: the next section surveys background literature on redistribution algorithms; Sect. 3 describes the proposed cycle redistribution algorithm; the experiments section compares the performance of the proposed algorithm variants using historical Transport For London data [14,17]. The final section discusses findings and draws conclusions.

## 2   Background

Multiple biology-inspired algorithms have been used for scheduling problems.

Ant Colony Optimization (ACO) is based on the foraging behaviour of ant colonies [3]. Artificial ants build solutions and exchange information through an indirect communication mechanism (stigmergy). An alternative method [10] proposes an implementation of Ant Colony System algorithm for a dynamic vehicle routing problem (VRP). The input is divided into multiple time segments, each is then treated as a static VRP. Their approach has three main components: an events manager to handle orders, Ant Colony System to solve a static VRP and most importantly a pheromone conservation matrix, used to store information on promising solutions such that it can be passed onto subsequent iterations.

The Genetic Algorithm (GA) is inspired by the concepts of natural selection and survival of the fittest [6]. There are many variations of GA's that make use of multiple populations through co-evolution, or that assemble smaller components of solutions together (e.g. classifier systems) [6]. Other approaches [11] propose a self organising method that applies a Genetic Algorithm to the redistribution of bicycle trucks for Santander Cycles (previously named Barclays Cycle scheme). Docking stations emit signals and trucks operate on these signals using local rules to rebalance the distribution of bicycles. A London Cycling Hire Index (LCHI) is maintained to measure the accessibility of different docking stations, a higher index indicates better accessibility and a station must have at least 2 bikes and 2 docking spaces in order to be accessible. Experiments showed that the proposed approach is preferable to random and greedy search algorithms. However their solution was proposed when the schemes scale was smaller in 2010; the dataset used had around 6000 bikes and 415 docking stations, which is 45 % smaller than the dataset used in this paper.

An alternative approach [11] proposes a routing algorithm based on greedy heuristics and a price incentive scheme for Santander Cycle scheme's customers Santander Cycle scheme. The price incentive scheme is used to reward customers who return their bikes to nearby less congested docking stations. Their approach first constructs a time-expanded network on a graph, a list of promising candidate solutions is built using a greedy heuristic, then the optimal number of bikes for each route is computed and the route with the highest utility is selected. A utility function is defined that estimates the value gained through a change in a stations state, this is achieved by computing the difference in the expected future usage of a docking station. This paper also defines a utility function, however our approach is different as we consider the status of docking stations.

## 3  Problem Description

### 3.1  Problem Overview

Currently Santander Cycle scheme [13,17] covers a $100\,\text{km}^2$ area of London with 748 docking stations strategically positioned and 11500 bikes, see Fig. 1. Table 1 introduces the key problem concepts.

The bike redistribution algorithm's goal is to maximise usage of the scheme by maintaining availability of bikes/spaces at docking stations. Too many bikes reduces the number of docking spaces available to return bikes, whilst not having enough bikes reduces the number of bikes available to commence a journey. Therefore there is an acceptable level of bikes, bounded by an upper and lower threshold, which must be maintained.

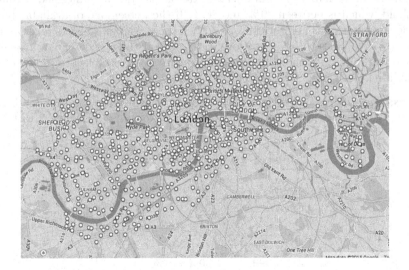

**Fig. 1.** TFL Santander Cycle Interactive Map showing the location of bike docking stations across London [15].

**Table 1.** Terminology

| | |
|---|---|
| Docking station | Terminal where bicycles are picked up or dropped off. Docking stations vary in capacity and usage level |
| Environment | Holds all docking stations |
| Job | Single pairwise movement of bikes from one station to another. A job is a solution building block, an individual member of a population |
| Schedule | Represents a sequence of jobs and their estimated operation time |
| Truck | A fleet of electric trucks, operated using a partially manual system, are used to redistribute bikes. All trucks have a fixed capacity of 18 bikes |
| Turnover | Percentage of jobs, where each job is an individual in the population, that is removed and replaced at the end of an iteration |
| Time segment | Time window during which we consider the problem to be static |

## 3.2 Performance Indicators

*Docking Station Status* a docking stations status depends on the percentage of available bikes. Docking stations have different capacities and therefore their status depends on the percentage of available bikes, see Table 2.

*Replenishment Level* is the number of bikes that must be added/removed for a docking station to maintain an acceptable status. The aim is to ensure the number of available bikes in docking stations is always maintained within a threshold, which is between 30 % and 70 %. If the percentage of available bikes in a docking station is greater than 70 % then the station has excess bikes and is eligible to give bikes to other stations. Otherwise if the percentage of available bikes is less than 30 % then this station needs to be given bikes to increase the overall level of service.

*Fitness Function* this fitness measure is used to evolve the solution, see Eq. 1. Where, $F_i$ = Fitness value for individual $i$. $F_{di}$ = Normalised Euclidean distance

**Table 2.** Docking station status

| Status | No bikes available |
|---|---|
| Empty | 0 % |
| Critical | $<= 10\%$ |
| Lower threshold | $<= 30\%$ |
| Acceptable | $30\% - 70\%$ |
| Upper threshold | $>= 70\%$ |
| Full | 100 % |

of the complete tour. $F_{ci}$ = Normalised schedule utility. $w_1$ = Weight for $F_{vi}$. $w_2$ = Weight for $F_{vi}$.

$$F_i = w_1 F_{di} + w_2 F_{ci} \tag{1}$$

**Nearest Neighbour Fitness** this fitness measure aims to minimise the distance between docking stations within a job and between jobs in a schedule. Where, $F_i$ = Fitness value for Job $i$. $F_{vi}$ = The Euclidean distance between docking stations. $F_{ci}$ = The Euclidean distance between stations in the Job. $w_1$ = Weight for $F_{vi}$. $w_2$ = Weight for $F_{vi}$.

$$F_i = w_1 F_{vi} + w_2 F_{ci} \tag{2}$$

**Schedule Time** the time taken to redistribute bikes is calculated using Eq. 3. Where, $T_i$ = time taken to perform a job $i$. $d_i$ = The Euclidean (Haversine formula) was used to calculate the distance between docking stations. $s_i$ = Speed (assumed to be 20 mph).

$$T_i = \frac{d_i}{s_i} \tag{3}$$

**Utility Function** is used to determine the value provided by a schedule, and therefore the benefit it provides to the scheme. A schedule's utility is calculated at each iteration, using Eq. 4. A higher utility implies that the schedule increases the scheme's service level. Where, $U(c_i)$ = Utility for a schedule $c_i$. $v_i$ = Value for Job. $S_n$ = Total number of stations.

$$U(c_i) = \frac{\sum v_i}{S_n} \tag{4}$$

A schedule is composed of a sequence of jobs. A job $j_i$ is associated with a value $v_i$ that can be positive or negative depending on the rules below. $S_{ic}$ is a docking stations current state. $S_{ip}$ is a docking stations predicted state. A docking station's status can be: $E$ = empty, $C$ = critical, $L$ = below lower threshold, $A$ = acceptable, $U$ = above upper threshold and $F$ = full.

1. $S_{ic} = A$ and $S_{ip} = E|C|L$ or $S_{ic} = A$ and $S_{ip} = F|U$ (negative)
2. $S_{ic} = E$ and $S_{ip} = C|L|A$ or $S_{ic} = F$ and $S_{ip} = U|A$ (positive)

### 3.3   Datasets

Experiments were performed using historical data from Transport For London library [14] and giCentre at City University London [12]. The data comprises:

1. Docking Station: A list of all the available docking stations in the scheme. Each station has the following parameters: a unique number identifier per docking station, location (latitude, longitude) and capacity docking points.
2. Historical TFL Dataset: This dataset captures the status of all docking stations, from TFL this data [12], at 10 min intervals. It was used to determine the historical service level of all the docking stations, which is indirectly representative of the service level provided by the scheme in operation.

3. Simulated TFL Dataset: A list of all successful journeys performed by users, also retrieved from TFL databank. It was used to simulate the journeys performed in the scheme and test the AEA.

# 4 Method

## 4.1 Assumptions and Restrictions

1. Trucks do not start and end at a depot. The start position is the first station in the schedule and the end position is the last station in the schedule.
2. The trucks speed is 20 mph at all times. Traffic flow is not considered.
3. In a schedule, a truck can only visit a station once. However, it is possible to visit the same station at the next time segment.
4. We do not consider the time it takes the truck driver to perform the redistribution.
5. The effect of weather, season, day of week, tube strikes.
6. The Euclidean formula is used to calculate the distance between docking stations.
7. Redistribution involves pairwise swaps between docking stations.

## 4.2 Bicycle Distribution Simulator

In order to assess the performance of bike distribution algorithms, we created a bicycle distribution simulator which uses the TFL dataset to simulate journeys from docks at different times, and calls a truck scheduling algorithm to redistribute bicycles as necessary. The proposed approach divides a day $d$ into a number of equally sized time segments $n$. This allows us to treat each time segment as a static problem. The idea of using time segments was originally proposed by [9]. It is possible to stop/start the optimiser when the problem changes, but this is unfeasible due to the large scale of the scheme, moreover this may prevent the optimiser from converging to a good solution [10]. Algorithm 1 shows the Bicycle Distribution Simulator.

## 4.3 Truck Scheduling Algorithms

This section describes five approaches used to dynamically redistribute bikes. The Random and Nearest Neighbour algorithms are first described, then three AEA variants are proposed: Artificial Ecosystem 1, Artificial Ecosystem 2 (using random) and Artificial Ecosystem 3 (using nearest neighbour).

**Random.** The Random Approach, see Algorithm 2, builds a schedule through stochastic selection of docking stations.

**Nearest Neighbour.** The Nearest Neighbour, see Algorithm 3, generates a schedule by selecting the nearest neighbouring docking station.

---

**Algorithm 1.** Bicycle Distribution Simulator

---

1: Split time period provided into days
2: Split day into time segments $T_i$
3: Get Station data $S_n$ from the *Historical TFL dataset*
4: **for each** Time Segment $T_i$ **do**
5:     Load *Journeys* performed from *Simulated TFL dataset*
6:     *Schedule* = Run Truck Scheduling Algorithm to obtain new schedule $S_i$
7:     Process *Journeys* and *Schedule* according to their timings
8: **end for**

---

**Algorithm 2.** Random

---

1: **loop**
2:     Pick a station at random $S_{i1}$
3:     If station has *excess* bikes: randomly select a station that *needs* bikes $S_{i2}$
4:     If station *needs* bikes: randomly select a station that has *excess* bikes $S_{i2}$
5:     Calculate the *Replenishment level* $R_i$
6:     Generate Job $J_i$ using $S_{i1}$, $S_{i2}$ and $R_i$
7:     Add $J_i$ to the *Schedule*
8: **until** *Schedule* reaches time segment limit

---

**Algorithm 3.** Nearest Neighbour

---

1: **loop**
2:     Pick a station at random $S_{i1}$
3:     If station has *excess* bikes: select a station that *needs* bikes $S_{i2}$ using Eq. 2
4:     If station *needs* bikes: select a station that has *excess* bikes $S_{i2}$ using Eq. 2
5:     Calculate the *Replenishment level* $R_i$
6:     Generate Job $J_i$ using $S_{i1}$, $S_{i2}$ and $R_i$
7:     Add $J_i$ to the *Schedule*
8: **until** *Schedule* reaches time segment limit

---

**Artificial Ecosystem 1.** The Artificial Ecosystem Algorithm (AEA) 1, see Algorithms 4 and 5, solves a problem by adapting subcomponents of a problem such that they fit together and form a single optimal solution. The problem is first decomposed into multiple sub-problems using a clustering algorithm, namely K-means++ or Self Organising Map (SOM). A population of jobs is created for each subproblem, then populations iteratively undergo a series of steps to form subsolutions in parallel. Firstly, a schedule is built using fitness based tournament selection, the schedule's utility is then evaluated and the best solution so far is updated, if necessary. Then fitness values of all jobs who have been part of the resulting schedule are updated using Eq. 1. Finally, a percentage of solutions with low fitness values are removed and new jobs are randomly created to replace them.

**Artificial Ecosystem 2 (Using Random Selection).** The Artificial Ecosystem Algorithm (AEA) 2, see Algorithms 4 and 5, first decomposes the problem

**Algorithm 4.** Artificial Ecosystem Algorithm

---

Initialise Environment $E$
**loop**
    Run SOM/K-means to build clusters of docking stations;
    **for each** Cluster $C_n$ **do**
        Create a population $P_n$ of Jobs;
        $Schedule = \textsc{ScheduleBuilder}(P_n)$
        Process $Schedule$;
    **end for**
    Connect all the segments to form a complete solution
    Evaluate overall solution
**until** All Time Segments processed

---

into multiple sub-parts using a clustering algorithm, then a population of jobs is created for each cluster. A solution is built by randomly selecting a sequence of jobs to form a schedule. The solution is then evaluated using the utility function, and the solution with the highest utility is used in subsequent stages of the simulation. At the end of each iteration, a percentage of solutions is removed and replaced randomly.

**Ecosystem Artificial 3 (Using Nearest Neighbour Selection).** The Artificial Ecosystem Algorithm (AEA) 3, see Algorithms 4 and 5, first decomposes the problem into multiple sub-problems using a clustering algorithm, a population of Jobs is created per cluster, then a schedule is generated by using the neighbour fitness equation Eq. 2. The best schedule has the highest utility and is used in later stages of the simulation. At the end of each iteration, jobs with a high Euclidean distance between stations are removed and replaced by randomly generated jobs.

## 5    Experiments

All experiments were repeated 50 times and results on the total number of full and empty docking stations, every time segment, each day, were recorded. The bike redistribution simulator and all the algorithms were built using Java. Whilst the Simulated and Historical TFL datasets were remotely held in a server running Mongo database instances. We ran experiments for 7 consecutive days, from 01/10/14 to 07/10/14. Data for all approaches was then gathered and compared in order to assess their relative performance:

1. Random.
2. Nearest Neighbour.
3. Artificial Ecosystem Algorithm (Random).
4. Artificial Ecosystem Algorithm (Nearest Neighbour).
5. Artificial Ecosystem Algorithm.

---

**Algorithm 5.** Schedule Builder

---

1: $MaxGen = 500$; $Iteration = 1$; $BestSchedule$;
2: **loop**
3:     **loop**
4:         Pick first Job at random;
5:         Use fitness based tournament selection to pick consecutive jobs;
6:     **until** Time constraint reached
7:     **if** Schedule utility $>$ Best Schedule utility **then**
8:         BestSchedule $=$ Schedule
9:     **end if**
10:     UPDATEFITNESS(SCHEDULE)
11:     REMOVEUNFITJOBS()
12:     CREATENEWJOBS()
13:     Iteration++;
14: **until** $MaxGen == Iteration$

---

6. Simulate TFL data, the same initialisation parameters given to the algorithms. To see what difference the algorithm does.

7. Historical data, the actual recorded status of the different docking stations.

## 6   Results and Analysis

The performance of all the docking stations for seven consecutive days, from Wednesday 01/10/14 to Tuesday 07/10/14 are compared in Fig. 2. The key performance indicator used, is the aggregate sum of full and empty stations at different time segments. The Simulated TFL and Random algorithm show very similar results. The three AEA's also show similar results, the fitness based AEA provides the best results for empty stations, whereas the Nearest Neighbour AEA provides the best results for full stations. These results show that the proposed ecosystem inspired algorithm performs promisingly, improving on or matching the results obtained from the Simulated and Historical TFL datasets.

The average performance of all the docking stations throughout the 7 consecutive days is presented in Fig. 3. The drop in the line on days 4 and 5, represents a reduction of user demand during the weekend, whereas during the week heavier use is recorded due to routine activities such as going to work. This graph clearly shows that the overall level of service provided by the Artificial Ecosystem Algorithm is an improvement based on the data gathered from the Simulated TFL and Historical TFL datasets.

The mean and standard deviation for all the implemented truck redistribution algorithms is given in Fig. 4. The results show that all implementations of the AEA perform well with low variance for both empty and full docks. This can be contrasted with the worse performance of the other approaches, and their higher variance for empty stations - not a desirable feature when consistency of service is important.

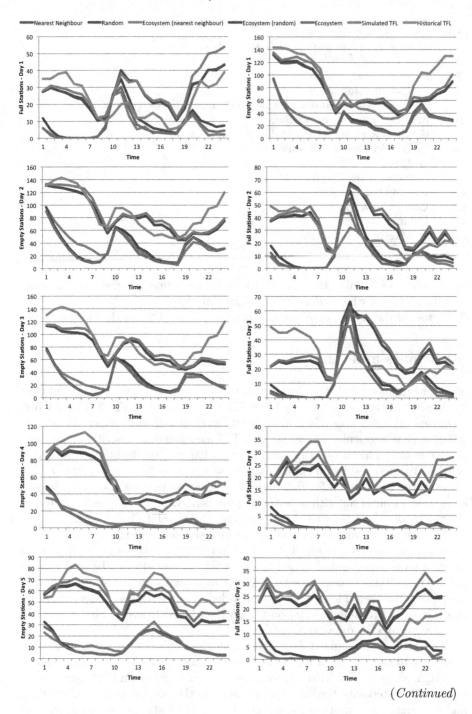

(Continued)

(*Continued*)

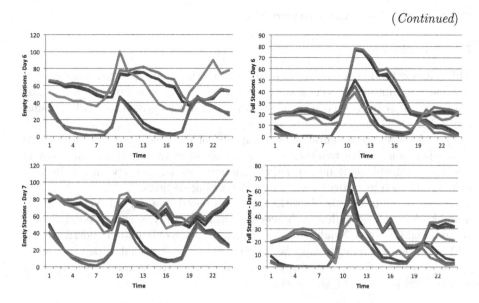

**Fig. 2.** Comparison of empty and full stations for the aforementioned datasets and algorithms across seven consecutive days (from 01/10/14 to 07/10/14)

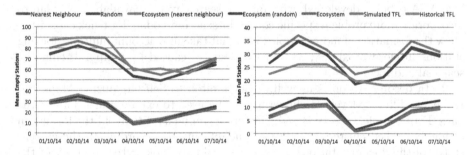

**Fig. 3.** Mean empty and full stations per day for each truck scheduling approach (low is good).

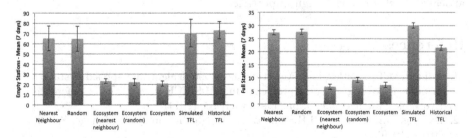

**Fig. 4.** Overall mean empty (left) and full (right) stations per truck distribution approach. Error bars show 1 standard deviation from the mean.

# 7 Conclusion and Future Work

## 7.1 Conclusion

This paper extended and applied the AEA to the problem of redistributing bikes in London's Santander Cycle scheme. The proposed approach divides a day into multiple time segments, each is then treated as a static problem and solved using the AEA.

The AEA shows potential, it is composed of multiple separate components that adapt and evolve together to form a complete system. Also, the AEA is easily configurable as it uses a utility function that can incorporate additional constraints such as priorities for different docking stations.

Three variants of this algorithm were proposed, namely AEA, AEA Random, AEA Nearest Neighbour. These algorithms were then compared against a Random and Nearest Neighbour algorithms, as well as Simulated and Historical TFL datasets. Computational results show that the Artificial Ecosystem algorithm reduces the number of empty and full docking stations, therefore improving service level provided by the scheme.

## 7.2 Future Work

The AEA is being developed, with many research avenues available. We plan to incorporate more problem specific constraints: accommodate for truck capacities, allow the formation of more complex jobs that take into account multiple docking stations in close proximity, as well as pressures caused by temporal, climatic, seasonal and local factors.

Data decomposition is as well as not limited to Self Organizing Maps and K-means, different levels of clustering can be used to expose dimensions of the data. For example, density Based Clustering can be used to avoid the formation of highly packed clusters, or we could adapt clusters according to the current status of the docking stations. It is also possible to look at the topology of the search space and select an appropriate decomposition method. Small clusters can be merged, large clusters can be decomposed, and more resources can be allocated to clusters with high importance.

# References

1. Adham, M.T., Bentley, P.J.: An artificial ecosystem algorithm applied to static and dynamic travelling salesman problems. In: 2014 IEEE International Conference on Evolvable Systems (ICES), pp. 149–156. IEEE (2014)
2. Dantzig, G.B., Ramser, J.H.: The truck dispatching problem. Manage. Sci. **6**(1), 80–91 (1959)
3. Dorigo, M., Maniezzo, V., Colorni, A.: Ant system: optimization by a colony of cooperating agents. IEEE Trans. Syst. Man Cybern. Part B: Cybern. **26**(1), 29–41 (1996)
4. Flood, M.M.: The traveling-salesman problem. Oper. Res. **4**(1), 61–75 (1956)

5. Gendreau, M., Potvin, J.-Y., Bräumlaysy, O., Hasle, G., Lokketangen, A.: Meta-heuristics for the vehicle routing problem and its extensions: a categorized bibliography. In: Golden, B., Raghavan, S., Wasil, E. (eds.) The Vehicle Routing Problem: Latest Advances and New Challenges, pp. 143–169. Springer, New York (2008)
6. Goldberg, D.E., Holland, J.H.: Genetic algorithms and machine learning. Mach. Learn. **3**(2), 95–99 (1988)
7. Hanshar, F.T., Ombuki-Berman, B.M.: Dynamic vehicle routing using genetic algorithms. Appl. Intell. **27**(1), 89–99 (2007)
8. Hu, D., Zhu, Z., Hu, Y.: Simulated annealing algorithm for vehicle routing problem. Chin. J. Highw. Transp. **4**, 022 (2006)
9. Kilby, P., Prosser, P., Shaw, P.: Dynamic VRPS: a study of scenarios. University of Strathclyde Technical report, pp. 1–11 (1998)
10. Montemanni, R., Gambardella, L.M., Rizzoli, A.E., Donati, A.V.: Ant colony system for a dynamic vehicle routing problem. J. Comb. Optim. **10**(4), 327–343 (2005)
11. Pfrommer, J., Warrington, J., Schildbach, G., Morari, M.: Dynamic vehicle redistribution and online price incentives in shared mobility systems (2013)
12. Slingsby, A.: Gicenter - City university (2015). http://www.gicentre.org/tfl_bikes/. Accessed 20 January 2015
13. tfl.gov.uk. London's Santander Cycle scheme (2015). https://www.tfl.gov.uk/info-for/media/press-releases/2015/february/mayor-announces-santander-as-new-cycle-hire-sponsor/. Accessed 20 April 2015
14. tfl.gov.uk. Open data users (2015). http://www.tfl.gov.uk/info-for/open-data-users/. Accessed 20 January 2015
15. tfl.gov.uk. TFL cycle hire map (2015). http://www.tfl.gov.uk/maps/cycle-hire/. Accessed 20 April 2015
16. Thangiah, S.R., Osman, I.H., Sun, T.: Hybrid genetic algorithm, simulated annealing and tabu search methods for vehicle routing problems with time windows. Computer Science Department, Slippery Rock University, Technical report SRU CpSc-TR-94-27, 69 (1994)
17. Wood, J., Slingsby, A., Dykes, J.: Visualizing the dynamics of london's bicycle-hire scheme. Cartographica: Int. Jo. Geogr. Inf. Geovisualization **46**(4), 239–251 (2011)

# Evolving Ensembles: What Can We Learn from Biological Mutualisms?

Michael A. Lones[1](✉), Stuart E. Lacy[2], and Stephen L. Smith[2]

[1] School of Mathematical and Computer Sciences,
Heriot-Watt University, Edinburgh EH14 4AS, UK
M.Lones@hw.ac.uk
[2] Department of Electronics, University of York,
York YO10 5DD, UK
{stuart.lacy,stephen.smith}@york.ac.uk

**Abstract.** Ensembles are groups of classifiers which cooperate in order to reach a decision. Conventionally, the members of an ensemble are trained sequentially, and typically independently, and are not brought together until the final stages of ensemble generation. In this paper, we discuss the potential benefits of training classifiers together, so that they learn to interact at an early stage of their development. As a potential mechanism for achieving this, we consider the biological concept of mutualism, whereby cooperation emerges over the course of biological evolution. We also discuss potential mechanisms for implementing this approach within an evolutionary algorithm context.

## 1 Introduction

Data mining is the process of finding meaningful patterns embedded within data. It typically comprises several stages, including feature extraction, feature selection, classification, and knowledge extraction. Here we focus on classification, which is the process of correctly identifying which category a previously unseen data point belongs to, based on the values of different explanatory variables (or *features*). Classification is the basis of computer-aided decision making, and has become an important tool in many spheres of knowledge, including biomedical diagnosis [9]. A classifier can be considered a function that maps features to classes. This function may be implemented in many different ways, including linear models, probabilistic models, decision trees, artificial neural networks, and support vector machines, to name but a few [5].

Traditionally, classification uses a single model instance, trained on a single sample of the data. Recently, however, there has been a shift towards ensemble models, which train multiple model instances (or base classifiers) and then combine their outputs [6]. By using a diverse selection of base classifiers, it is often possible to reach more accurate and robust decisions than those produced by individual classifiers. Diversity amongst the base classifiers is typically achieved by training them on different samples of the data and/or by restricting them to using different subsets of the features. Widely used examples include bagging, boosting and random forests [5,6].

© Springer International Publishing Switzerland 2015
M. Lones et al. (Eds.): IPCAT 2015, LNCS 9303, pp. 52–60, 2015.
DOI: 10.1007/978-3-319-23108-2_5

A common element of ensemble generation algorithms is that the base classifiers are generated sequentially. With the notable exception of boosting, where the performance of each classifier influences the training of the next one, base classifiers are also typically trained independently. This is in stark opposition to the patterns of interaction seen within biological systems, where the elements of cooperative systems emerge in parallel and with considerable inter-dependency. This begs the question: is there some advantage to the biological approach? Moreover, can we learn how to build better ensembles by studying the organisation and evolution of biological systems? To answer this question, we begin by surveying some of the previous work on using models of evolution, namely evolutionary algorithms, to generate ensembles. We also survey previous work on introducing ideas of biological mutualism to evolutionary algorithms. We then discuss how mutualistic models might be practically applied to the problem of ensemble generation, outlining potentially useful research directions.

## 2 Evolving Ensembles

Evolutionary algorithms (EAs) are population-based optimisation algorithms modelled upon the process of biological evolution. From an ensemble classification perspective, an evolutionary algorithm's population is a potentially valuable resource. During the course of evolution it often contains multiple solutions with diverse behaviours, which is precisely what we are looking for in an ensemble. This has promoted interest in using EAs to train ensemble classifiers.

However, in general, an EA's population loses diversity as evolution progresses. This means that it is often necessary to use a diversity preservation mechanism to maintain behavioural diversity. In our work, we have been looking at whether niching methods can be used to maintain a diverse population of base classifiers throughout an EA run, and whether we can then form effective classifiers from the final population [7,8,13]. Niching techniques in EAs are motivated by observations from biology, notably the appearance and preservation of ecological niches within evolving biological populations. Common examples are crowding, fitness sharing and geographical segregation. Our results are promising, showing that the resulting ensembles are significantly more predictive than individually trained classifiers [8]. More recently, we have applied this approach to a medical diagnosis problem, finding that evolved ensembles can compensate for the small differences in clinical practice that often occur between different sites in medical studies. Because evolved ensembles comprise a relatively small number of individually strong classifiers, rather than many weak classifiers (as in random forests), there is also the potential to extract knowledge, which is important for diagnostic classifiers used within a medical context [10].

In biology, niching often reflects the loss of competition between organisms that have no need to compete, for instance because they use different energy and nutrient sources. Species in different niches can then cooperate [16], producing mutualistic interactions that can be beneficial to the ecosystem as a whole [5]. Interestingly, a classifier ensemble is in some ways similar to an ecosystem. It comprises a group of different classifiers (akin to species) that reach their decisions (their behaviour) as a result of processing different data sources. In general, it makes no sense for classifiers in different niches to compete. However, it does make sense for them to cooperate. We might therefore expect that biological mutualism can guide us in finding more effective ways of evolving ensembles.

## 3   Biological Mutualism

A mutualism is a symbiotic relationship that is beneficial to all parties [19,20]. These are very common in biology. For example, over 80 % of plant species form mycorrhizal associations with fungi, in which the fungus absorbs sugars from the plant's roots and the plant receives phosphates and other hard to absorb nutrients that the fungus scavenges from the surrounding area [21]. This allows the plant to grow in nutrient-poor environments and the fungus to grow in energy-poor environments. Other well known examples include mutualisms between flowering plants and pollinating insects, and between mammals and their colonies of gut bacteria. In many cases, evolution has driven the process to the stage where the individual members of the mutualism can not survive without their mutualistic partners. Mutualistic interactions can involve more than two species [2]. Furthermore, wider interactions involving mutualistic partners lead to the formation of mutualistic networks, which may involve hundreds of species, and which have been described as the 'architecture of biodiversity' [1]. It has also been suggested that selection inevitably pushes populations in the direction of ever increasing cooperation and integration, so the scope and connectivity of these mutualistic networks is likely to increase over evolutionary time [19].

An important issue, both for biology and for any system that seeks to emulate biology, is understanding the conditions required for mutualisms to occur. In particular, what are the mechanisms that prevent the emergence of selfish behaviour in a population of cooperating species—or, to put it another way, which mechanisms support collective action? This question is also studied in the context of economic and social interactions, where it is framed as dealing with asymmetric or hidden information, i.e. where there is a lack of trustworthy information regarding a potential partner's intentions or abilities. Two broad mechanisms for dealing with this situation are signalling and screening [20]. Signalling involves an up-front expenditure of energy or resources (a *strategic waste*) in order to attract cooperation from a potential partner. Examples are the investment in costly phenotypes (e.g. large tail feathers) in animal mating displays, and the distribution of dividends by companies looking for investors. Screening is similar, but is initiated by the less well-informed partner; for example, the expectation that potential suitors should fight for mating rights in order to demonstrate their fitness, or requiring an employee to undergo a probationary period. In symmetric situations, where neither party has sufficient knowledge of the other, combinations of both signalling and screening are often used.

# 4  Mutualistic Evolutionary Algorithms

EAs are based upon relatively simple models of biological evolution. Ideas of biological mutualism, which have become increasingly influential in the understanding of biological systems, have so far had little impact upon their development. Nevertheless, there are notable exceptions, and several forms of EA do exhibit behaviours that resemble those seen in biological mutualisms.

Co-evolutionary algorithms [14], in particular, model the co-evolution of multiple species within one or more populations. Although mutualistic relationships are co-evolutionary, the two terms are not synonymous, and co-evolution can also lead to symbiotic, predator-prey, constructive or destructive relationships between species. Co-evolutionary algorithms are mainly restricted to two-species systems, and model either competitive (e.g. predator-prey) or cooperative co-evolutionary systems. The latter approach has been particularly successful in recent years, with cooperative co-evolutionary algorithms being successfully used to solve high-dimensional optimisation problems [22]. We have also had some success using these within the context of ensemble classification. For instance, in [13], this mechanism was used to evolve pattern recognisers (i.e. classifiers) in DNA sequences, using a second population of pattern combinators (i.e. ensembles) to promote the discovery of meaningful groups of patterns. This was only possible because the base classifiers were evolved in parallel, using information about their ensemble behaviour to focus the search towards the most useful population of base classifiers.

Although not directly motivated by biological mutualism or co-evolution, two other forms of EA that display behaviours akin to mutualism are Michigan-style learning classifier systems and cultural algorithms. Learning classifier systems (LCS) [3] evolve distributed rule sets that can be used to determine responses, for example class predictions in classification problems. In a Michigan-style LCS, each member of the population is a simple rule; however, the overall behaviour of the LCS is a result of cooperation between the rules. Cultural algorithms [15] use a shared belief space where population members can read and write information. This provides a mechanism for population members to interact with one another, and this in turn can be used to support mutualism, amongst many other kinds of behaviour. Both systems use mechanisms to discourage selfish behaviour. In the case of the LCS, rewards are divided between groups of interacting rules using credit assignment techniques such as the *bucket brigade*. Selfish behaviour by a rule will reduce the overall reward for the group, reducing the likelihood of it being propagated to the next generation. In the case of cultural algorithms, only the fittest population members are typically allowed to write to the belief space, with the expectation that fitter members are more likely to be part of a beneficial mutualism. A notable feature of LCS and certain cultural algorithms [18] is that the population *is* the solution, since individual members of the population may not have a meaningful function when separated from other population members. In this respect, they bear significant resemblance to mutualistic biological systems. However, the mechanisms used to promote mutualisms are quite different to those seen within biological systems.

## 5   Evolving Mutualistic Ensembles

So, might biological ideas of mutualism be useful for evolving ensembles? The combination of specialism and co-operation seen within biological populations is precisely the behaviour we desire in classifier ensembles. We want individual base classifiers to focus on particular sub-tasks, such as processing different subsets of the data (comparable to different nutrient and energy sources). We then want them to co-operate, combining their behaviours in a way that increases their mutual fitness, i.e. we want them to be well-adapted to one another. With current ensemble methods, the behaviours of base classifiers are combined in a variety of simple ways, such as voting or averaging of their outputs (Exceptions include our work on evolving non-linear output combinators [7]). In addition, the ensembles produced are strictly hierarchical, with only the raw features delivered to the inputs, and only the final outputs delivered to the combinator.

It would be interesting to explore forms of interaction more akin to those we see in biology. For example, the output of one classifier could be delivered to an intermediate decision node within another classifier. This is more comparable to a plant/fungus mutualism, with a classifier that is specialised at processing a difficult part of the decision space (analogous to sparse nutrients) feeding into a classifier that makes broader judgements. Using suitable classifier models, we could even remove the hierarchical organisation of classifiers entirely, allowing cyclic interactions to occur (see Fig. 1). Recurrent dynamical systems models, such as recurrent ANNs, and artificial biochemical networks [11], would be appropriate for this. Our work on artificial signalling networks [4] is particularly relevant, showing how complex decision making networks can be formed from interactions between simpler recurrent pathways. However, these are just some of the possible models of representation and interaction that could be

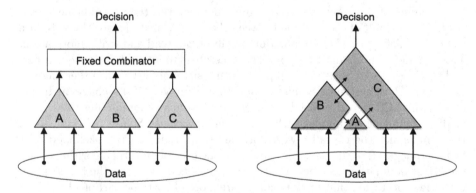

**Fig. 1.** Left — the standard form for an ensemble classifier, with information feeding hierarchically upwards from the data set to the individual classifiers to the combinator. Right — a mutualistic arrangement in which information flows between classifiers. Classifier A is not able to reach a decision by itself, but generates an important sub-decision that feeds into the rest of the system. An appropriate form of credit assignment would then cause it to be promoted in the next generation.

used within a mutualistic classifier ensemble, and there is considerable scope for adventure in this area. Nevertheless, all approaches would have to address the issue of how to promote the emergence of mutuality in classifier populations.

## 5.1  Promoting Mutuality

Biological organisms enter into mutualisms because it directly improves their fitness. A similar motivation is required for classifiers, otherwise there is no incentive for them to enter into mutualistic partnerships. In general, we expect to achieve this by redistributing the fitness of the ensemble amongst its component base classifiers, with the expectation that ensembles will outperform individual classifiers and therefore exert selective pressure. Reinforcement learning techniques, such as those used in LCS, could be used for this redistribution. However, there are potential complexities. For example, it would not be unreasonable for a single classifier to be involved in multiple mutualistic ensembles. In such a circumstance, should it receive a reward from all its partnerships, or just the most successful one? The former might promote overly-promiscuous classifiers, yet the latter might discourage wider interactions. Also, should a classifier involved in an ensemble also receive its standalone fitness? This would promote the propagation of individually strong classifiers, but would disadvantage a classifier that plays an important role in supporting an ensemble but which has little standalone worth.

Credit allocation acts as a carrot for the formation of mutualistic interactions, but a stick is also needed to prevent selfish or inefficient behaviours akin to parasitic or symbiotic relationships in biology. Based on biological understanding, signalling and screening have the potential to be effective mechanisms for achieving this. For instance, classifiers could be required to pay a fitness penalty in order to be involved in an ensemble. This form of signalling would be expected to discourage selfish behaviour, since parasitic classifiers would lose individual fitness at the same time as reducing the potential fitness payback from joining an ensemble. There is also the potential for using such a system to study biological hypotheses of how mutualisms are formed. Unlike many of the game theoretic models typically used for this, an evolving ensemble has its own complex behavioural outcomes. In this respect, the mutualistic evolution of classifier ensembles can be seen as an opportunity to test, refine, and potentially even discover biological theories of mutualism.

## 5.2  Identifying Partners

Biological mutualisms involve different forms of interaction. Mycorrhizal associations between plants and fungi, for instance, usually involve direct contact, with the fungus identifying and then invading the cells of its interacting plant species. Other kinds of mutualism are based upon geographical proximity. For example, a tree provides shelter for birds, birds excrete beside the tree, and the tree absorbs the nutrients. This is also an example of using the environment (the soil in this case) as an intermediary for sharing nutrients, energy, or information.

The classifiers involved in an ensemble must also be able to identify one another, and must be able to do this in a way that is robust to evolutionary change. For instance, a classifier could reference another classifier by a pointer. However, pointers are not robust to evolution, since the reproduction process usually involves cloning, and often leads to multiple offspring. An alternative is to use the environment as a go-between. LCS, for example, do this, with the behaviour of a rule changing the environment, and this change of context then leading to the firing of another rule. It is also the basis of interaction in cultural algorithms, with information being transferred via the belief space environment. However, environmental interaction can lead to unintended interactions, and can increase the scope for selfish behaviour.

Another alternative is to use some form of functional referencing in which the reference describes the form or behaviour of the intended partner. Assuming that no major change in form or behaviour takes place during reproduction, this form of referencing would also be fairly robust to evolution. An example of a functional reference is our work on implicit context representation in genetic programming systems [12].

## 6   Conclusions

Biological systems are a potentially useful source of information and motivation for the design of computer algorithms. However, there is a danger of taking inappropriate motivation from biology, or of unthinkingly applying bio-inspired algorithms to inappropriate problems. This has led to a degree of criticism in the literature [17]. In this paper, we propose that inspiration from biological mutualisms could be applied to the design and generation of classifier ensembles. To us, this is an example of where biological systems display a clear example of how to achieve complex behaviour in a system of distributed decision makers, and where useful lessons could be learnt and applied to a problem which shows parallels to these biological systems. We have discussed some of the issues that would be involved in doing this. However, we have not tried to suggest one particular way of doing this, because there are many paths that could be taken. Rather, we wish to begin a discussion of the ways in which this could be done, and the potential benefits to which it could lead.

Whilst we have restricted our attention to the particular problem of generating classifier ensembles, it seems likely that mutualism could also provide solutions to other kinds of problem addressed using EAs. For example, genetic programming is concerned with evolving executable structures, such as computer programs. In this context, mutualistic interactions could potentially be used to address the scalability barrier associated with evolving single large programs. Mutualisms could even develop over a series of evolutionary runs, with subsequent runs interacting with solutions found in earlier runs, and possibly bringing them back into the evolutionary process. We would argue that mutualism is an important area of biology that is often overlooked by the evolutionary computation community, but which could bring significant benefits to this community if modelled and used appropriately. Doing so would also provide an opportunity to test, and potentially generate, biological theories of the ways in which mutualisms evolve.

**Acknowledgements.** This work was supported by the EPSRC [grant ref. EP/M013677/1].

# References

1. Bascompte, J., Jordano, P.: Plant-animal mutualistic networks: the architecture of biodiversity. Annu. Rev. Ecol. Evol. Syst. **38**(1), 567–593 (2007)
2. Biere, A., Bennett, A.E.: Three-way interactions between plants, microbes and insects. Funct. Ecol. **27**(3), 567–573 (2013)
3. Bull, L.: Learning Classifier Systems: A Brief Introduction. Applications of Learning Classifier Systems, pp. 1–12. Springer, Berlin (2004)
4. Fuente, L.A., Lones, M.A., Turner, A.P., Stepney, S., Caves, L.S., Tyrrell, A.M.: Computational models of signalling networks for non-linear control. BioSystems **112**(2), 122–130 (2013)
5. Kuhn, M., Johnson, K.: Applied Predictive Modeling. Springer, Berlin (2013)
6. Kuncheva, L.I.: Combining Pattern Classifiers. Wiley-Interscience, Chichester (2004)
7. Lacy, S., Lones, M.A., Smith, S.L.: A comparison of evolved linear and non-linear ensemble vote aggregators. In: Proceedings of 2015 Congress on Evolutionary Computation, CEC 2015. IEEE Press, May 2015
8. Lacy, S., Lones, M.A., Smith, S.L.: Forming classifier ensembles with multimodal evolutionary algorithms. In: Proceeding of 2015 Congress on Evolutionary Computation, CEC 2015. IEEE Press, May 2015
9. Lones, M.A., Smith, S.L., Alty, J.E., Lacy, S.E., Possin, K.L., Jamieson, D.R.S., Tyrrell, A.M.: Evolving classifiers to recognize the movement characteristics of parkinson's disease patients. IEEE Trans. Evol. Comput. **18**(4), 559–576 (2014)
10. Lones, M., Alty, J.E., Lacy, S.E., Jamieson, D., Possin, K.L., Schuff, N., Smith, S.L., et al.: Evolving classifiers to inform clinical assessment of parkinson's disease. In: 2013 IEEE Symposium on Computational Intelligence in Healthcare and e-health (CICARE), pp. 76–82. IEEE (2013)
11. Lones, M.A., Turner, A.P., Fuente, L.A., Stepney, S., Caves, L.S., Tyrrell, A.M.: Biochemical connectionism. Nat. Comput. **12**(4), 453–472 (2013)
12. Lones, M.A., Tyrrell, A.M.: Modelling biological evolvability: implicit context and variation filtering in enzyme genetic programming. BioSystems **76**(13), 229–238 (2004)
13. Lones, M.A., Tyrrell, A.M.: A co-evolutionary framework for regulatory motif discovery. In: IEEE Conference on Evolutionary Computation, CEC 2007, pp. 3894–3901. IEEE (2007)
14. Popovici, E., Bucci, A., Wiegand, R.P., De Jong, E.D.: Coevolutionary Principles. Handbook of Natural Computing, pp. 987–1033. Springer, Berlin (2012)
15. Reynolds, R.G.: An introduction to cultural algorithms. In: Proceedings of the Third Annual Conference on Evolutionary Programming, pp. 131–139 (1994)
16. Santos, F.C., Pinheiro, F.L., Lenaerts, T., Pacheco, J.M.: The role of diversity in the evolution of cooperation. J. Theor. Biol. **299**, 88–96 (2012)

17. Sörensen, K.: Metaheuristics—the metaphor exposed. Int. Trans. Oper. Res. **22**(1), 3–18 (2015)
18. Spector, L., Luke, S.: Cultural transmission of information in genetic programming. In: Proceedings of the First Annual Conference on Genetic Programming, pp. 209–214. MIT Press (1996)
19. Stewart, J.E.: The direction of evolution: the rise of cooperative organization. Biosystems **123**, 27–36 (2014)
20. Turcotte, M.M., Corrin, M.S.C., Johnson, M.T.J.: Adaptive evolution in ecological communities. PLoS Biol. **10**(5), e1001332 (2012)
21. Wang, B., Qiu, Y.L.: Phylogenetic distribution and evolution of mycorrhizas in land plants. Mycorrhiza **16**(5), 299–363 (2006)
22. Yang, Z., Tang, K., Yao, X.: Large scale evolutionary optimization using cooperative coevolution. Inf. Sci. **178**(15), 2985–2999 (2008)

# An Artificial Immune System for Self-Healing in Swarm Robotic Systems

Amelia R. Ismail[1], Jan D. Bjerknes[2], Jon Timmis[3]([⊠]), and Alan Winfield[4]

[1] Department of Computer Science, International Islamic University Malaysia,
P.O. Box 10, 50728 Kuala Lumpur, Malaysia
`amelia@iium.edu.my`
[2] Kongsberg Defence Systems, P.O. Box 1003, 3601 Kongsberg, Norway
`jan.dyre.bjerknes@kongsberg.com`
[3] York Robotics Lab and Department of Electronics, University of York,
Heslington, York, UK
`jon.timmis@york.ac.uk`
[4] Bristol Robotics Lab, University of the West of England, Bristol, UK
`Alan.Winfield@uwe.ac.uk`

**Abstract.** Swarm robotics is concerned with the decentralised coordination of multiple robots having only limited communication and interaction abilities. Although fault tolerance and robustness to individual robot failures have often been used to justify the use of swarm robotic systems, recent studies have shown that swarm robotic systems are susceptible to certain types of failure. In this paper we propose an approach to *self-healing* swarm robotic systems and take inspiration from the process of granuloma formation, a process of containment and repair found in the immune system. We use a case study of a swarm performing team work where previous works have demonstrated that partially failed robots have the most detrimental effect on overall swarm behaviour. In response this, we have developed an immune inspired approach that permits the recovery from certain failure modes during operation of the swarm, overcoming issues that effect swarm behaviour associated with partially failed robots.

## 1 Introduction

Swarm robotics is an approach to the co-ordination and organisation of multi-robot systems of relatively simple robots [6]. Traditional multi-robot systems employ centralised or hierarchical control and communication systems in order to coordinate behaviours of the robots. Swarm robotics, however, adopts a decentralised approach, in which the desired collective behaviours emerge from the local interactions and communications between robots and their environment. Work in [6] argues that a significant benefit of swarm robotics is robustness to failure. However, recent work has shown, that for certain modes of operation, swarm robotic systems are not as robust as first thought [2,3]. Work in [3] proposed a simple, but effective, algorithm, the $\omega$-algorithm, for emergent swarm taxis (swarm motion towards a beacon) under sensory

© Springer International Publishing Switzerland 2015
M. Lones et al. (Eds.): IPCAT 2015, LNCS 9303, pp. 61–74, 2015.
DOI: 10.1007/978-3-319-23108-2_6

constraint. In order to achieve beacon-taxis, the algorithm provides coherence through simple rules of operation, and introduces a simple symmetry breaking mechanism which permits the emergence of beacon-taxis. To understand the reliability of the $\omega$-algorithm, work in [2,3] undertook an evaluation of the effect of individual robot failures on the operation of the overall swarm. Potential failures investigated were: (1) complete failure of individual robots due to, for example, a power failure (2) failure of a robot's IR sensor and (3) failures of robot's motors only, leaving all other functions operational including the sensing and signalling. The study revealed that the effect of motor failures will have a potentially serious effect of causing the partially-failed robot to 'anchor' the swarm impeding its movement towards the beacon.

Work described in this paper uses the $\omega$-algorithm and investigates a failure mode specific to motor failures due to a lack of power in the robots: a variation on (3) outlined above. We detail an immune-inspired solution which enables, under the given failure mode, the swarm to 'self-heal' through simulated *trophallaxis*, the exchange of power between units, to permit continued operation and overcome 'anchoring' of the swarm. In order to develop the immune-inspired algorithm, we adopt an immune-engineering approach, as outlined in [10]. Specifically, we take inspiration from the process of granuloma formation in the immune system: a process of 'containment and repair' observed during various inflammatory responses. We develop a simple agent-based simulation, from which we derive a set of design principles that are used to design an algorithm capable of isolating the effect of the failure and initiate a recovery strategy. We explore the performance of the proposed algorithm, in simulation, and show that an effective self-healing system can be operationalised within the $\omega$-algorithm, thus increasing tolerance to failure of the swarm.

The rest of the paper is structured as follows. In Sect. 2 we provide the necessary background to the $\omega$-algorithm and discuss the issues of 'anchoring' of the swarm. In Sect. 3 we provide an overview of initial experimental investigations into the $\omega$-algorithm that were undertaken for this work and provide a baseline of results against which to compare our proposed approach. In Sect. 4 we introduce our immune-inspired approach, a granuloma-formation inspired system. In this section we provide a basic introduction to the underlying biology and four design principles that were abstracted from an agent-based simulation, developed by the authors, that captured the basic granuloma operation. In Sect. 5 we provide simulation based experimental results from our investigations and conclude in Sect. 6.

## 2   Robotic Swarm Taxis

Aggregation of a swarm requires that agents maintain physical coherence when performing a task. Robots are placed in an environment, ensuring that they are within signalling distance of each other, and interact with each other to maintain swarm coherence. This is relatively easy when a centralised control approach is used, but very challenging when a distributed control approach is used [1].

Work in [2] and [3] developed a class of aggregation algorithms which makes use of local wireless connectivity information alone to achieve swarm aggregation namely the $\alpha$, $\beta$ and $\omega$ algorithms.

For work in this paper, we make use of the $\omega$-algorithm which contains two swarm behaviours: flocking and swarm taxis towards a beacon. This combination means that the swarm maintains itself as a single coherent group, whilst moving toward an infra-red (IR) beacon. The algorithm is a modified version of the wireless connected swarming algorithm (the $\alpha$-algorithm) developed by [5]. The wireless communication channel in $\omega$-algorithm has been removed and replaced with simple sensors and a timing mechanism. Flocking is achieved through a combination of attraction and repulsion mechanisms. Repulsion between robots is achieved using IR sensors and a simple obstacle avoidance behaviour. Attraction is achieved using a simple timing mechanism. Each robot measures the duration since its last avoidance behaviour. If that time exceeds a threshold, then the robot turns towards its own estimate of where the center of the swarm is. It will move in that direction until it, once again, must avoid the other robots in the swarm. In order for the swarm to reach the beacon, the algorithm uses a symmetry-breaking mechanism, in which the short-range avoid sensor radius for those robots that are illuminated by the beacon is set slightly larger than the avoid sensor radius for those robots in the shadow of other robots [3]. An emergent property of this approach is swarm taxis towards the beacon.

## 2.1 Anchoring of the Swarm

Various types of failure modes and the effect of individual robots failures and its effect to the swarm have been analysed by [2]. From [2] the failure modes and effects for swarm beacon taxis are as follows:

- Case 1: complete failures of individual robots (completely failed robots due, for instance, to a power failure) might have the effect of slowing down the swarm taxis toward the beacon. These are relatively benign, in the sense that 'dead' robots simply become obstacles in the environment to be avoided by the other robots of the swarm.
- Case 2: failure of a robot IR sensors. This could conceivably result in the robot leaving the swarm and becoming lost. Such a robot would become a moving obstacle to the rest of the swarm and might reduce the number of robots available for team work.
- Case 3: failure of a robot's motors only. Complete motor failure only leaving all other functions operational, including IR sensing and signalling. Such a failure will have the potentially serious effect of causing the partially-failed robot to 'anchor' the swarm, impeding its taxis toward the beacon.

The effect of completely failed robot(s) is to 'anchor' the swarm thus impeding its taxis toward the beacon. If in the swarm there are only one or two robots that are subject to complete failure, the swarm will still move towards the beacon. This is a form of a *self-repairing* mechanism inherent in the $\omega$-algorithm.

With complete failure of a robot, the 'dead' robots simply become obstacles in the environment to be avoided by the other robots of the swarm, thus temporarily slowing the swarm down. However, under a partial failure mode, the swarm will experience a serious effect which causes the partially failed robot to 'anchor' the swarm. In swarm beacon taxis, this can only happen if the *anchoring force* resulted by the effect of the robot's motor failure are greater than the *beacon force*, which is the force pulling the swarm toward the beacon.

A certain number of robots are necessary to maintain the emergent swarm taxis property. A reliability model (k-out-of-n-system model) of the swarm in swarm beacon taxis has been developed and the results show that there is a point at which the swarm no longer functions [2]. The results from that paper suggest that in a swarm of 10, then at least 5 robots have to be working in order for swarm taxis to emerge [2]. This would indicate that in order for the swarm to continue operation then some form of *self-healing* mechanism is required apart from *self-repairing* mechanism which is already available in swarm beacon taxis [3].

Work in this paper will consider the failure mode in which there is not enough power to drive the motors, but sufficient for signalling via LEDs on the robots. This is classified as a partial failure. This assumption has been tested electronically. We performed a simple experiment with epuck robots where we allowed the robots to wander in an environment with a simple obstacle avoidance behaviour until the robots stopped moving. We monitored the power levels within the robots from this point (when the robots stopped moving) and when the battery was totally discharged. We found that on average, the e-puck robots are able to send signals for 27 min before all the energy is lost.

The following section investigates the $\omega$-algorithm [3] in a Player/Stage simulation. This will serve as a baseline against which we will be able to compare our proposed immune-inspired algorithm.

## 3    Initial Investigations into the $\omega$-algorithm

### 3.1    Experimental Protocol

The experiments presented in this section were performed in simulation using the sensor-based simulation tool set, Player/Stage [4][1]. 10 e-puck robots are simulated, sized $5\,cm \times 5\,cm$, and equipped with 8 proximity sensors, two at the front, two at the rear, two at left and two at right. Initially robots are dispersed within a 2 m circle arena with random headings, ensuring that IR communication between robots was possible. A robot will poll its proximity sensors at frequency $5\,Hz$ $(1/T)$, whenever one or more sensors are triggered the robot will execute an avoidance behaviour, and turn away from the colliding robot or obstacles. The avoidance turn speed depends on which sensors are triggered and robots will keep turning for 1 s. The task of the swarm is to aggregate and move together towards an infra-red beacon located in the arena. The parameters for the simulation are provided in Table 1. Each simulation run consisted of ten robots and was

---

[1] Player-Stage can be downloaded from http://playerstage.sourceforge.net/.

**Table 1.** Robot fixed parameters for all simulations.

| Parameter | Value |
|---|---|
| Time step duration | 1 s |
| Robot normal speed | 15 cm/s |
| Avoidance sensor range | 4 cm |
| Robot body radius | 12 cm |
| Energy left in battery in order to move | 500 J |
| Battery capacity | 5000 J |
| Component fault | power drain |
| No of faulty robots | 1 to 5 units |
| Simulation duration | 1000 s |

repeated ten times. For each run, the centroid position of the robots in swarm were recorded.

As outlined above, the failure mode for our experiments is a motor failure, which is due to low battery power. We assume that the failure is sufficient to stop the motors of the robot, but that there is sufficient power to light LEDs. Using the simulation, we inject a power reduction failure to robots: for a single robot failure, for two robot failures and three or more robot failures (until we reach five failures) which will be introduced simultaneously in the simulation. When the robots fail, they are not moving and will remain static in the environment. The parameters for the failed robots in this scenario are shown in Table 2.

In these experiments we measure the progression of the centroid of the swarm towards the beacon for every 100 s, using Eq. 1; where $x$ and $y$ are the coordinates of the robots and $n$ is the number of robots in the experiment and $cd$ is the centroid distance of robots to beacon. Statistical tests were performed using the Mann-Whitney rank sum test, effect magnitude tests were performed using the A-Test.

$$cd = \sum_{i=1}^{n} \frac{\sqrt{(x_{1_i} - x_{2_i})^2 + (y_{1_i} - y_{2_i})^2}}{n} \tag{1}$$

We undertook a series of experiments on the $\omega$-algorithm to reproduce work in [3]. For the sake of space we report only a single hypothesis tested, which formed part of work by [3] and allows us to establish a baseline of performance

**Table 2.** Variable parameters for failing scenario in the environment

| Number of faults | Parameter | Time (s) |
|---|---|---|
| Single failure | Speed = 0 m/s, energy = 500 joules | t = [100] |
| Multiple failure | Speed = 0 m/s, energy = 500 joules | t = [100] |

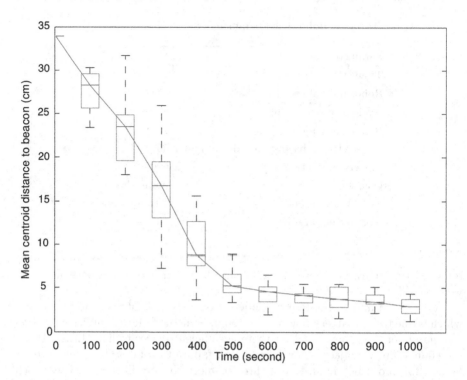

**Fig. 1.** Boxplots of the distance between swarm centroid and beacon as a function of time for 10 experiments using $\omega$-algorithm with no robot failure for $H1_0$. The centre line of the box is the median while the upper edge of the box is the $3^{rd}$ quartile and the lower edge of the box is the $1^{st}$ quartile. At about $t = 600$ seconds, the swarm has reached the beacon

for the $\omega$-algorithm: $H1_0$: *The $\omega$-algorithm for swarm beacon taxis allows the swarm to achieve a centroid distance less than 5 cm away from the beacon when there are no failures introduced.*

The swarm starts the experiment in one part of the arena and begins to move toward the beacon. The distance from the centroid of the swarm to the beacon for each run is given in Fig. 1. For each experiment the robots have a different starting position in the arena, but as the importance here is on the relative performance between different sets of runs, the starting point was set to 35 cm from the beacon. This allows for a comparison between each run, as comparison between runs will be for identical starting distances from the beacon. The hypotheses can be accepted if the swarm reaches a distance of less than 5 cm from the beacon. Based on the experiments, the swarm has a mean velocity of 1.2 mm/simulation seconds. The fastest moved at 1.52 mm/simulation seconds and the slowest had the velocity of 1.01 mm/simulation seconds. At time $t = 600$ seconds, the swarm has reached 5 cm from the beacon.

Now we have established a baseline of operation with respect to the algorithm, and reproduced previous work, we now proceed to detail our proposed immune-inspired solution. Experimental results reproducing the 'anchoring' problem can be found in Sect. 5.

## 4   Immune Inspiration: Granuloma Formation

Of particular interest to our work is the formation of structures known as *granuloma* in response to certain pathogenic infection. Granulomas form in response to cells being infected by pathogens (in particular in response to infection by tuberculosis and Leishmania). They are structures that form of particular immune cells, known as T-cells that try and contain the infection from spreading inside the host [8]. The main cells involved in granuloma formation are macrophages, T-cells and cytokines. Work by [8] outlines three main stages of granuloma formation:

1. T-cells are primed by antigen presenting cells;
2. Cytokines and chemokines are released by macrophages, activated T-cells and dendritic cells. The released cytokines and chemokines attract and retain specific cell populations.
3. The stable and dynamic accumulation of cells lead to the formation of the organised structure of the granuloma.

These stages do vary slightly depending on location of the granuloma, be that in the liver or lung, for example, but the principles are common.

The Leishmania parasite infects a cell known as a Kupfer cell. If the cells begin to break down (cell lysis) these cells will begin to release chemokines, notably IL-10, which in turn beings to attract macrophages and T-Cells to form around the infected cell, in an attempt to "contain" the infected cell. This process will ultimately lead to the formation of a granuloma, and in many cases the resolution of the infection. By the creation of the granuloma structure, resolution from the infection is possible, however in some cases, the infection is fatal. For the purposes of our work, we consider the most important property in granuloma formation to be the communication between cells, and the recruitment of T-cells, which is determined by the level of chemokine secretion (IL-10). Chemokines will not only attract other macrophages to move towards the site of infection but will activate T-cells that will secrete cytokines to act as a signal for activation of macrophages. T-cells and activated macrophages are able to kill extra cellular bacteria that will control infections in a host.

Using the conceptual framework described in [9] for the development of immunologically grounded algorithms, we prepared a simple agent based simulation to provide the framework to allow us to understand how the process of granuloma formation occurs in the immune system. The agent based simulation can provide insight into the behavioural aspects of a granuloma formation, which can then be abstracted into design principles for algorithm development. Therefore, by modelling and simulating the properties of granuloma formation we can attempt to formalise principles that govern the behaviour of cells in the system and apply them towards the development of a solution in our specific engineering problem. Through the analysis of our developed model and simulation, we have constructed four design principles of self-healing in swarm robotic systems. The four principles for our algorithm development are:

1. Communication between agents in the system is indirect (so not direct from one specific robot to another), and should consist of a number of signals that allow for coordination of the agents.
2. Agents in the systems react to defined failure modes in a self-organising manner.
3. Agents must be able to learn and adapt by changing their role dynamically.
4. Agents can initiate a self-healing process dependant to their ability and location.

We now move to the development of a *granuloma inspired* approach to a self-healing swarm.

### 4.1 Granuloma Formation Algorithm

Using our knowledge gained from the development of a computational model, and the subsequent derivation of the design principles outlined above, we have derived a granuloma formation algorithm for addressing the *anchoring issue* in swarm beacon taxis. The algorithm is outlined in Algorithm 1. Failing robots will send signals that can be recognised by other functional robots. These functional robots are then attracted towards the faulty robot, akin to how T-cells are attracted by cytokines emitted by an infected macrophage in granuloma formation. A limited number of these robots then isolate the faulty robot. The other robots which are not involved in isolating the faulty robot will ignore the failed and surrounding robots and treat them as if they were obstacles in a manner similar to the standard $\omega$-algorithm.

> **begin**
>> Deployment: robots are deployed in the environment ;
>> **repeat**
>>> Random movement of the robot in the environment;
>>> Communication: Faulty robots will get the information on distance and energy of its neighbour, where the average radius of the target neighbouring robot is $R$;
>>> ;
>>> Protection and rescue: Healthy robots will decide how many will perform protection and rescue according to Algorithm 2;
>>> Repair: Sharing of energy between faulty and healthy robots according to Algorithm 3;
>> **until** *forever*;
> **end**

**Algorithm 1.** Overview of Granuloma Formation Algorithm

In our proposed algorithm, the number of functional robots that will come to the aid of a faulty robot varies, it is not pre-defined and therefore is *dynamic*.

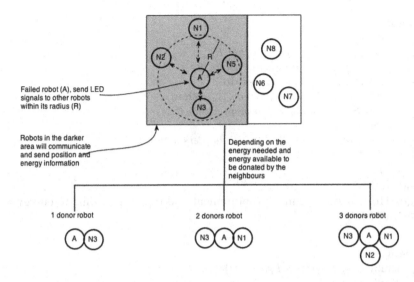

**Fig. 2.** The number of functional robots that will share their energy will be based on its position, their current energy and the energy needed from the faulty robot.

The number of robots required will be determined by the amount of energy required to repair the failed robot, together with the location of the faulty robot. Each faulty robot is able to evaluate their own energy level and position, and propagate that information to other robots, as outlined in Algorithm 2. An illustration of this process is shown in Fig. 2. When a faulty robot(s) are present, they will emit a signal to other robots, which can only receive that signal if they are within a certain predefined radius (R). The robots that receive the signal will communicate and exchange information on their current battery energy levels, in accordance with the energy transfer rules detailed in Algorithm 3. The nearest functional robot will attach to the failed robot and share an amount of energy, ensuring that the robot providing the energy does not deplete their own resource to such a degree as to enter a failure mode itself. If sufficient energy can be donated, the robot being repaired will stop emitting the energy request signal. However, if the energy is not sufficient, the faulty robot will continue to request energy from other functional robots in the environment. This will be repeated until enough energy has been transferred to the robot to permit resumed operation.

The basic terms used in the algorithm are as follow:

- $pos_{self}(t)$: position of self robot
- $pos_{peer}(x)$: position of peer robots
- $egy_{self}(t)$: energy of self robot
- $egy_{peer}(x)$: energy of peer robots
- $egy_{needed}$: energy needed by failing robot
- $egy_{threshold}$: the limit of the energy that is needed

**begin**

    Evaluate $egy_{needed}(t)$ ;

    Send $egy_{needed}(t)$ to peers within R;

    Receive $egy_{peer}(x)$ from peers within R;

    **forall the** $egy_{peer}(x)$ received **do**

        **if** $egy_{peer}(t)$ received $< egy_{needed}$ **then**

          | Send $egy_{needed}(t)$ to peers within R;

        **else**

          | Stop sending $egy_{needed}(t)$ to peers within R;

        **end**

    **end**

**end**

**Algorithm 2.** Algorithm for containment and repair according to energy and position of robots

**begin**

    Evaluate $egy_{self}(t)$ and $pos_{self}(t)$;

    Send $egy_{self}(t)$ and $pos_{self}(t)$ to peers within R;

    Receive $egy_{peer}(x)$ and $pos_{peer}(x)$ from peers;

    **forall the** $egy_{peer}(x)$ received **do**

        **if** $egy_{peer}(x) < egy_{min}$ **then**

          | $R = match\ egy_{peer}(x),\ egy_{self}(t)$ ;

          | Store $egy_{peer}(x)$ in inbound queue ;

        **else**

          | $R' = not\ match\ egy_{peer}(x),\ egy_{self}(t)$;

          | Add $egy_{peer}(x)$ to outbound queue;

        **end**

    **end**

    **forall the** $egy_{peer}(x)$ in inbound queue **do**

        Add signature ;

        Store $egy_{peer}(x)$ in robot list

        **forall the** $egy_{peer}(x)$ in robot list **do**

          **if** $egy_{self}(t) < egy_{threshold}$ **then**

            | Evaluate $pos_{peer}(x)$ ;

            | Sort $pos_{peer}(x)$ in ascending order;

            | Move to nearest $pos_{peer}(x)$ ;

          **end**

        **end**

    **end**

    **forall the** $egy_{peer}(x)$ in outbound queue **do**

        | Delete signature ;

    **end**

**end**

**Algorithm 3.** Algorithm for containment and repair according to energy and position of robots

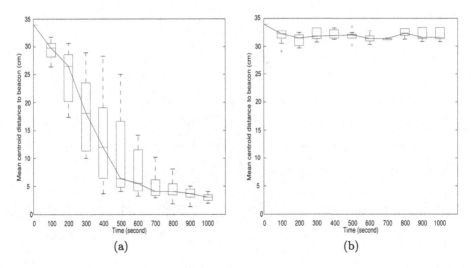

**Fig. 3.** Boxplots of the of the distance between swarm centroid and beacon as a function of time for 10 experiments using the $\omega$-algorithm, one failing robot (Fig. 3(a)) and three failing robots (Fig. 3(b))

For a full hardware deployment, during the repair phase, donor robots will be able to dock share energy with the faulty robot, using a platform such as [7].

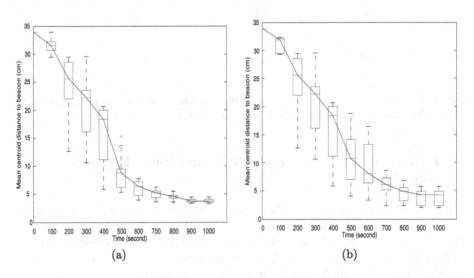

**Fig. 4.** Boxplots of the distance between swarm centroid and beacon as a function of time for 10 experiments using granuloma formation algorithm with one faulty robot (Fig. 4(a)), and three failing robots (Fig. 4(b))

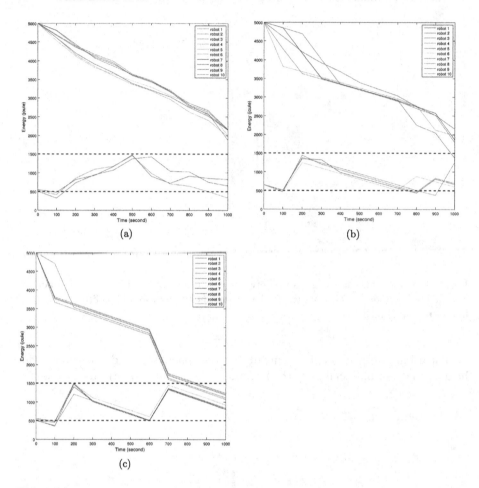

**Fig. 5.** The energy autonomy for swarm of 10 robots using granuloma formation algorithm with three (Fig. 5(a)), four (Fig. 5(b)) and five (Fig. 5(c)) failing robots.

## 5   Experimental Investigation: Granuloma Formation Algorithm

For this experiment, we follow the experimental protocol that is described in Sect. 3.1. We cast our hypothesis as follows:

$H2_0$: *The use of a granuloma formation algorithm does not improve the ability of the robots in the system to reach the beacon when compared to the $\omega$-algorithm when more than two faulty robots are present in the swarm.*

### 5.1   Results

We compared the performance of the $\omega$-algorithm when one and then three faults were introduced, this is shown in Fig. 3(a) and (b). As can be seen progress is made to the beacon with a single fault (Fig. 3(a)) but when three faults are present, the anchoring problem is clearly shown and the mean centroid distance does not change (Fig. 3(b)). However, when the granuloma-inspired system was tested with three failing robots, we can see from Fig. 4(a) that swarm taxis is still successful (when compared to Fig. 3(a)) and importantly when tested for $H2_0$ we can see that from Fig. 4(b) we are unable to accept the hypothesis, as even in the presence of three failures, swarm taxis is successful in permitting the swarm to reach the beacon. We were able to reject the hypothesis at the 5% confidence level.

### 5.2 Maintaining Energy Across the Swarm

The use of energy across the swarm can also be investigated. As in previous experiments, all robots start with equal energy which is 5000 Joules. Figure 5(a), (b) and (c) show the energy of 10 robots during 1000 simulation seconds with different failing robots in the environment with granuloma formation algorithm. Having three to five failing robots, we can see that the average energy level of all robots in the environment are still above the minimum energy level which is 500 Joules. This shows that the robots can share their energy even though half of the robots are experiencing low energy levels. The swarm is able to continue operating even with such failures.

## 6  Conclusion

Our work has proposed a novel immune-inspired system, a granuloma formation algorithm for self-healing mechanism in swarm robotic systems applied in a case study of swarm beacon taxis, specifically the $\omega$-algorithm developed by [3]. Our experiments show that the granuloma formation algorithm is able to provide a mechanism for 'self-healing' under certain failure models, being able to initiate a simple recovery strategy through the recruitment of capable robots in the surrounding area, which are able to perform trophollaxis and transfer energy between one robot and another. Further work on this approach would demand the demonstration on a suitable hardware platform to demonstrate the applicability away from simulation only.

**Acknowledgements.** The authors would like to thank International Islamic University Malaysia and Ministry of Higher Education of Malaysia in funding this research, the CoSMoS project, funded by EPSRC grants EP/E053505/1 and EP/E049419/1 and the SYMBRION project under FP7: FET Proactive Initiative: PERVASIVE ADAPTATION (PERADA), grant number FP7-ICT-216342. JT is part funded by The Royal Academy of Engineering and The Royal Society.

## References

1. Bayindir, L., Şahin, E.: A review of studies in swarm robotics. Turk. J. Electr. Eng. Comput. Sci. **15**(2), 115–147 (2007)

2. Bjerknes, J.D., Winfield, A.F.T.: On fault tolerance and scalability of swarm robotic systems. In: Martinoli, A., Mondada, F., Correll, N., Mermoud, G., Egerstedt, M., Hsieh, M.A., Parker, L.E., Støy, K. (eds.) Distributed Autonomous Robotic Systems. STAR, vol. 83, pp. 431–444. Springer, Heidelberg (2013)

3. Bjerknes, J.D.: Scaling and fault tolerance in self-organized swarms of mobile robots. Ph.D. thesis, University of the West of England, Bristol, UK (2009)

4. Gerkey, B., Vaughan, R.T., Howard, A.: The player/stage project: tools for multi-robot and distributed sensor systems. In: ICAR 2003, pp. 317–323 (2003)

5. Nembrini, J., Winfield, A., Melhuish, C.: Minimalist coherent swarming of wireless networked autonomous mobile robots. In: Proceedings of the 7th International Conference on Simulation of Adaptive Behavior (SAB 2002), vol. 7, pp. 373–382. MIT Press (2002)

6. Şahin, E.: Swarm robotics: from sources of inspiration to domains of application. In: Şahin, E., Spears, W.M. (eds.) Swarm Robotics 2004. LNCS, vol. 3342, pp. 10–20. Springer, Heidelberg (2005)

7. Schlachter, F., Meister, E., Kernbach, S., Levi, P.: Evolve-ability of the robot platform in the symbrion project. In: 2008 Second IEEE International Conference on Self-Adaptive and Self-Organizing Systems Workshops, SASOW 2008, pp. 144–149, October 2008

8. Sneller, M.C.M.: Granuloma formation, implications for the pathogenesis of vasculitis. Clevel. Clin. J. Med. **69**(2), 40–43 (2002)

9. Stepney, S., Smith, R., Timmis, J., Tyrell, A., Neal, M.J., Hone, A.N.W.: Conceptual framework for artificial immune systems. Unconv. Comput. **1**(3), 315–338 (2005)

10. Timmis, J., Andrews, P., Hart, E.: On artificial immune systems and swarm intelligence. Swarm Intell. **4**, 247–273 (2010)

# Team Search Tactics Through Multi-Agent HyperNEAT

John Reeder[(✉)]

SPAWAR Systems Center Pacific, San Diego, USA
john.d.reeder@navy.mil

**Abstract.** User defined tactics for teams of unmanned systems can be brittle and difficult to define. The state and action space grows with each new system added to the team which increases the difficultly in designing robust behaviors. In this paper we present a method for using Multi-agent HyperNEAT to develop tactics for a team of simulated unmanned systems that is robust to novel situations, and scales with the number of team members. We focus on the tactics of a search area coverage task, where the need for team work, and robust asset management are critical to success.

## 1   Introduction

The use of remote operated and autonomous vehicles is increasing as the size and cost of the unmanned platforms continues to decrease. Drones are being used by farmers to survey crops, photographers for aerial views, and volunteer search and rescue groups. With the advent of cheap consumer friendly drones like the Parrot ARDrone, and the DJI Phantom, it is now possible to build low cost swarms of drones. As the number of vehicles grows the requirement for each to have an individual operator will become ever more burdensome. To alleviate this, more robust autonomy needs to be developed, especially in the case of large numbers of drones acting in concert, such as a swarm.

The specific case of a team of unmanned systems can be treated as a multi-agent system (MAS). Multi-agent systems is a very active area of research. It is interesting because of the joint behaviors and the complexities arising from the interactions of agents with some degree of autonomy [13]. Early work in MAS dealt with planning and scheduling [12], and had very little to do with learning. It was primarily focused on developing protocols for interaction, and studying the effect of various levels of agent communications. Early methods used scripted or rule based controls [9] or symbolic systems [6,8]. However, because of the increased complexity of designing and building MAS, machine learning quickly came into play. For an in depth survey of Machine Learning in MAS see [15,21] or [13].

In this paper, we will explore the use of Multi-Agent HyperNEAT [7] in developing search tactics for a team of autonomous agents. Tactics, in this case, refer to the collective search paths of the team. The team will need to work together to provide continuous coverage of designated search areas. MA-HyperNEAT has

© Springer International Publishing Switzerland 2015
M. Lones et al. (Eds.): IPCAT 2015, LNCS 9303, pp. 75–89, 2015.
DOI: 10.1007/978-3-319-23108-2_7

been used in developing team behaviors in the predator-prey problem as well as for teams of robots attempting to completely compounds tasks requiring teamwork [14]. In this work we are interested in how the incorporation of tactical information about the environment impacts the tactics developed through the evolutionary process.

The following sections will go into the background of NEAT, HyperNEAT, and MA-HyperNEAT, followed by a description of our experiments and finally our results.

## 2   Background

### 2.1   NeuroEvolution of Augmenting Topologies (NEAT)

*Neuro-evolution* (NE; [27]) is a technique for training artificial neural networks (ANNs) through evolutionary algorithms. Instead of updating connection weights through a learning rule like back-propagation, ANNs are generated through evolutionary processes, evaluated in a task and given a fitness that used to select and create a new generation of ANNs through mutations and recombination of their genomes. The NeuroEvolution of Augmenting Topologies (NEAT) algorithm [20] is a popular neuroevolutionary approach that has been proven in a variety of challenging tasks, including particle physics [1, 26], simulated car racing [2], RoboCup Keepaway [22], function approximation [25], and real-time agent evolution [17], among others [20].

NEAT begins with a population of minimal size, simple ANNs which increase in complexity over generations through adding new nodes and connections using mutation. This allows the topology of the network to evolve over time rather than being defined a priori. In this way NEAT searches through increasingly complex networks to find the appropriate level of complexity. The techniques used to evolve a population of increasingly complex and diverse networks are described in [20]. The major concept that should be noted for this paper is that NEAT is a successful method that discovers the best topology and weights for a neural network that maximizes the performance on a task. NEAT has been extended to allow for indirect encodings which allows it to evolving very large networks with a method called HyperNEAT. This technique is detailed in the next section.

### 2.2   CPPNs and HyperNEAT

Hypercube-based NEAT (HyperNEAT; [10,18]) is an extension of NEAT that allows for the evolution of high-dimensional ANNs. The effectiveness of the HyperNEAT algorithm has been demonstrated in multiple domains; multi-agent predator prey [4,5] and RoboCup Keep away [24]. The full description of Hyper-NEAT is available in [18].

The core idea in HyperNEAT is that geometric relationships are learned through an indirect encoding that describes how the *weights* of the ANN can be *generated* as a function of geometry. In a standard direct representation each connection weight is represented as part of the genome, whereas an indirect representation describes a pattern of parameters without enumerating each individual parameter. This allows information to be reused which is a major focus in the field of Generative and Developmental systems from which HyperNEAT originates [19,23]. This information reuse allows indirect encodings to search a compressed space. HyperNEAT discovers the regularities in the geometry of a problem and learns from them.

The indirect encoding in HyperNEAT is called a *compositional pattern producing network* (CPPN; [16]), which encodes the *weight pattern* of an ANN [11,18]. The idea behind CPPNs is that geometric patterns can be encoded by a *composition of functions* that are chosen to represent common regularities. Given a function $f$ and a function $g$, a composition is defined as $f \circ g(x) = f(g(x))$. In this way, a set of simple functions can be composed into more elaborate functions through hierarchical composition. (e.g. $f \circ g(f(x) + g(x))$). Formally, CPPNs are *functions* of geometry (i.e. locations in space) that output connectivity patterns for nodes situated in $n$ dimensions. Consider a CPPN that takes four inputs labeled $x_1$, $y_1$, $x_2$, and $y_2$; this point in four-dimensional space can *also* denote the connection between the two-dimensional points $(x_1, y_1)$ and $(x_2, y_2)$. The output of the CPPN for that input thereby represents the weight of that connection (Fig. 1). By querying every pair of points in the space, the CPPN can produce an ANN, wherein each queried point is the position of a neuron. While CPPNs are themselves networks, the distinction in terminology between CPPN and ANN is important for explicative purposes because in HyperNEAT, CPPNs *encode* ANNs.

Because the connection weights are produced as a function of their endpoints, the final pattern is produced with *knowledge* of the domain geometry, which is

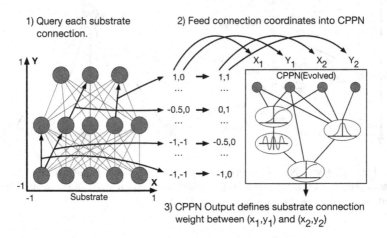

**Fig. 1.** The CPPN encodes the values of the connection weights between any two nodes in the substrate.

literally depicted geometrically within the constellation of nodes. Weight patterns produced by a CPPN in this way are called *substrates* so that they can be verbally distinguished from the CPPN itself. It is important to note that the structure of the substrate is independent of the structure of the CPPN. The substrate is an ANN whose nodes are situated in a coordinate system, while the CPPN defines the connectivity among the nodes of the ANN. The experimenter defines both the location and role (i.e. hidden, input, or output) of each node in the substrate. As a rule of thumb, nodes are placed on the substrate to reflect the geometry of the domain (i.e. the state), making setup straightforward [10,18].

## 2.3   Multi-Agent HyperNEAT

HyperNEAT is extended further in [5] to allow the algorithm to produce networks for multiple agents. It introduces the idea of *policy geometry* which is the concept that the behaviors of a team are a continuum between *homogeneous* and *heterogeneous* extremes. If every member of a team were completely heterogeneous then there would be no shared skills amongst the team mates. MA-HyperNEAT is able to generate a spectrum of policies by extending the HyperNEAT idea to a new dimension. In MA-HyperNEAT each connection is a function of its nodes end points and the position of its parent network along the spectrum of the policy geometry. Using a $z$-stack parameter MA-HyperNEAT is able to generate a near infinite amount of policies that share common traits but that differ according to their position in the team. It is this ability in MA-HyperNEAT that we will take advantage of in our work. Figure 2 shows a simple CPPN with the $z$ input and a hypothetical stack of generated networks.

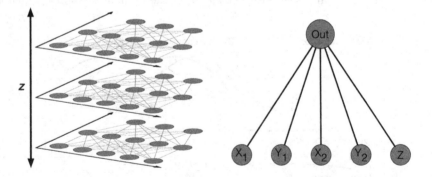

**Fig. 2.** The MA-HyperNEAT algorithm uses a $z$ parameter to adjust the position of the agents policy. This allows MA-HyperNEAT to create a stack of networks that have different behaviors based on its position in the team.

# 3   Experimental Method

This section will detail the simulator and the experiments in learning team search tactics. The primary focus of these experiments is to determine which combinations of sensors and control schemes allow MA-HyperNEAT to develop effective search tactics. The learned tactics are compared to a selection of hand coded tactics scripted at the team level.

## 3.1   Search Task

The experiments in this paper are designed to explore the ability of MA-HyperNEAT to develop team tactics, and how the representation of the environment and agents effects the development of these tactics. The task explored in these is experiments is a cooperative search task where the agents work together to provide continuous coverage of designated search areas. Tactics in this scenario refer to the collective search patterns followed by the team of agents. Each agent has a limited area they can cover and they must work together to maximize their combined coverage.

Figure 3 shows the simulated search environment used in each of the experiments. There are two separate search areas to provide the opportunity for the team to split up and cover multiple areas. The experiments are carried out with a team of 3 simulated agents. This team size was chosen to minimize processing time for the experiments while still allowing for MA-HyperNEAT to learn multiple team roles. The size and placement of the search areas are fixed between experiments so that the coverage score calculation is based solely on the effectiveness of the search patterns.

To evaluate how well each search tactic covers the area, and to provide a fitness function for the evolutionary process we use a measure of decayed coverage. Equation 1 provides the details for the *DecayedCoverage* measure. This measures the average search area covered by the team over time. The measure decays at an exponential rate of about 5 % per time step. In order to maximize this measure the team must move quickly and minimize the overlap between their sensor coverage. This measure is used to provide an indication of how well the team maintains continuous coverage of the search areas. Since speed has a large influence over this measure, and the intended goal of the evolution is interesting search patterns, the speed of the evolved agents, and the scripted agents are held to the same maximum velocity. This is done to make the comparisons fair. With speed held at an identical level the values become primarily dependent on minimizing the overlap between the agents sensors and traversing over unseen search area.

$$DecayedCoverage = \frac{\sum_{a=0}^{A} \sum_{x,y=0}^{X,Y} V_{x,y}(t)}{t} \quad (1)$$

$$V_{x,y}(t) = \begin{cases} e^{-t/20} & t \text{ time steps since last view} \\ 0 & >20 \text{ time steps since last view} \end{cases} \quad (2)$$

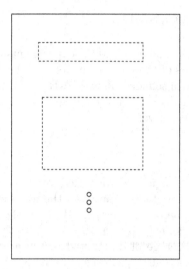

**Fig. 3.** Sample search environment. The boxes defined by the dotted lines indicate areas that need to be searched.

## 3.2　Agent Simulator

The experiments were carried out using the MA-HyperNEAT Agent simulator [3] developed by the EPLEX group at UCF. The simulator models a 2D environment and includes collision detection, first order kinematics, and several object types. Experiments can be setup to include any number of simulated agents, points of interests, start points, goal points, obstacles, and walls. The simulator is built to interface with the HyperSharpNEAT code base which is included in the download with several pre-configured experiments, robots types, and sensor types included as well.

In order to implement the multi-agent search experiment it was necessary to extend the simulator in several ways. To implement search areas a new simulation object type was added based on the existing area of interest. The area of interest is a special rectangle in the simulation that bounds the environment. Walls are placed at the edges to keep the agents from leaving the area. The search areas are also represented in the simulation by special rectangles, but with additional properties that allow them to interact with the agent sensors, and to track where the agents have already searched. On the agent side several new sensor types were added to allow the agents to sense the search areas both at a lower relative level, and at a higher tactical level. These sensors are used in conjunction with the existing sensors in the simulator to give the agents their view of the environment.

## 3.3　Sensing Tactical Environments

The primary goal of this research is to learn tactics for a team of agents. The theory is that in order for the MA-HyperNEAT algorithm to be able to develop tactics, it needs to be able to incorporate tactical information. In this instance tactical information means a higher level overview of the environment. In previous multi-agent experiments [7] MA-HyperNEAT has been used to develop networks for each individual agent in a predator-prey scenario. In these experiments each agent has its own relative sensors and controls. This means each agent's interactions are based solely on it's local environment. The sensors described below provide the inputs to the neural networks developed by the MA-HyperNEAT algorithm, while the type of control scheme determines the output of the networks. The combination of sensors and controls determines the structure of the neural network substrate.

The standard relative sensors used in these experiments are radar sensors, and ray sensors. Radar sensors detect the presence of other simulation objects (search areas, other agents) within a specified range, and arc around the agent. The radar segments surround the agent providing a 360° field of coverage. The value returned by the radar sensor depends on what is being sensed. For search areas the radar returns a value from 0 to 1 representing the percentage of the radar cone that is unseen search area. For other agents the value represents the distance to the detected agent again in the range of 0 to 1. In these experiments 8 radar segments are used. The ray sensors detect intersections along a particular relative angle from the agent. The ray sensors are binary so the value represents a detection or no detection. Ray sensors are used to sense the edge of the environment. The ray sensors are not used by the agent control network, they are used by the simulation to turn the agents away from the environment boundaries. Figure 4 shows a graphical depiction of a radar segment and an array of ray sensors.

To incorporate tactical information into the agents behavior a new sensor type is added to compliment the local sensors of the agents. Figure 6(a) shows a representation of a grid sensor. The grid sensor overlays the entire environment and provides the agent with two types of information. The first type of information is the absolute location of all of the search areas. The second type of information is the absolute position of all of the agents in the environment including itself. In our experiments we explore several combinations of these inputs. The GridHeading input combines the grid sensor with a relative heading sensor. This setup is intended to give the individual agents an indication of

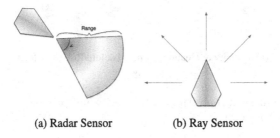

(a) **Radar Sensor**           (b) **Ray Sensor**

**Fig. 4.** The Radar sensors provide a 360° coverage around the agent, while the ray sensors provide binary detection along specific relative angles

their heading in addition to their absolute position. The RadarGrid input is a combination of the search area radar and the grid sensor. These combinations are tested in addition to the grid, and radar sensors individually.

The standard controls tested in these experiments are velocity and heading (VelHeading), grid heading (GridOut), and absolute heading (AbsOut). Velocity and heading uses 3 output signals to determine the movement of the agent. Two of the values are used to determine the heading, one output represents left, while the other represents right. The difference in their output signals determines the change in heading of the agent. If the left output activates stronger than the right, then the agent will turn towards the left. The size of the change is determined by the size of the difference. The velocity output determines the speed value from 0 to max speed. The grid heading control uses 8 output signals and represents the cardinal positions of a compass. Each output signal is polled and the highest value is chosen, the agent then moves in the direction represented by that output. The speed of the agent depends on the value of the output signal, a max output signal corresponds to max speed. The absolute heading output is similar to the grid heading scheme; the difference is the relative heading represents 8 directions in relation to the agent; forward, left, right, backward, and the angles in between.

In addition to adding the grid sensor to provide high level tactical input, a new type of output is added as well. Figure 6(b) shows a vector field output (VectorOut). This type of output is designed to give a high level sense of direction to the agent. The standard outputs for the agents are a velocity and relative heading which are poled at each time step to determine the agents next course of action. The vector field on the other hand provides a heading vector for each point on the grid corresponding to the grid sensor. At each time step the agents position is polled and its current heading vector is moved a fraction of the way toward the vector field desired heading.

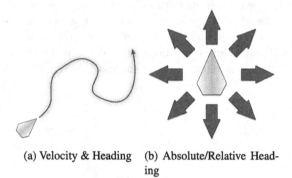

(a) Velocity & Heading    (b) Absolute/Relative Heading

**Fig. 5.** The standard agent control schemes. Velocity and heading use 3 output signals, while the two heading controls use 8.

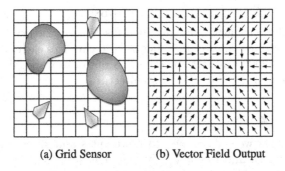

(a) Grid Sensor          (b) Vector Field Output

**Fig. 6.** The Grid Sensor (a) gives a tactical overview of the environment while the Vector Field Output (b) gives a high level policy that applies a directional force on the agents heading.

### 3.4 MA-HyperNEAT Agents

To develop team tactics the MA-HyperNEAT algorithm is used to generate neural network controllers for each agent in the team. The search tactics for the evolved controllers result from the interaction of these individual agent behaviors. The agents interact with each other through their sensors of the environment and do not have explicit communication. The MA-HyperNEAT algorithm has been shown to develop cooperative behaviors without the need for explicit communication in previous work [7].

The choice of sensors and control scheme determines the size and shape of the neural substrate for each agent. MA-HyperNEAT works by evolving a CPPN to set the weights of the fixed substrate. In our experiments the substrates are designed to match the different input/output choices of each experiment. In the Grid/Vector Field cases the grids are set to $20 \times 20$ grids. The substrates are laid out so that the input/output nodes follow a 2D pattern matching the grid sensor or the vector field out. For the radar and heading inputs each sensor has eight sub divisions leading to the substrate having 8 input nodes for each of these sensors. The VelHeading output uses three nodes; one for the velocity value and two to indicate the relative heading (left or right). The GridOut and AbsOut outputs each have eight subdivisions meaning eight nodes in the substrate for each of those outputs. The hidden layers of the substrate used two primary shapes. For the inputs involving grids the hidden layer was set to a $10 \times 10$ grid, while the radar only substrates used a hidden layer of 20 nodes.

These dimensions for the substrates lead to very large neural networks. The largest network arises from the Grid input/Vector Field output. Each agent has two input grids of 400 input nodes each (one for the search areas, and one for agent locations), a hidden layer of 100 nodes, and an output layer of 400 nodes. The substrates are fully connected so this leads to a network with 120000 connections. Each agent has their own substrate so the total number of connections for the experiment totals 360000. Using the standard NEAT or other neuro-evolution techniques would mean evolving with these 360000 parameters, but the indirect encoding of these weights using the evolved CPPN of MA-HyperNEAT allows us to evolve new solutions much more quickly. This is one of the primary advantages of the MA-HyperNEAT approach. Using other techniques each additional agent would introduce 120000 more connections to optimize while MA-HyperNEAT only has to evolve the smaller CPPN architecture.

### 3.5   Scripted Agents

The scripted agents were developed to represent standard search tactics used in patrols and search and rescue operations. The scripted agents have perfect information about the locations of the search areas and thus do not make use of the sensors in the simulation environment. This allowed the scripted agents to be explicitly programmed for the desired patterns, rather than having to develop rules and behaviors based on sensor inputs to elicit the patterns. Three distinct patterns were programmed into the scripted agents. The ladder pattern or lawnmower pattern moves up and down the area moving over slightly after each pass. The spiral pattern moves along the outside of the search area in a cycle and then moves inward at the end of each cycle. The random pattern simply chooses a new random direction once it reaches the edge of a search area.

The scripted agents are controlled by a single monolithic controller. The controller takes the position of each agent into account and sends their next control signals based on where they are in the desired pattern. The controller also handles load balancing by dividing the search areas up into equal portions based on the number of agents available to search; each agent gets their own sub area to patrol.

### 3.6   Evolution and Evaluation

In order for the MA-HyperNEAT algorithm to develop the team controllers each team must be evaluated on the search task. The CPPNs that develop the best teams continue on to the next generation during the evolutionary process. Each of the evaluation runs starts with the three agents below the search areas lined up behind a start point. The agents are evaluated for 300 simulated seconds, and their *DecayedCoverage* score at the end of the run is reported back to the MA-HyperNEAT algorithm. The evolution continues for 2000 generations and the CPPN that developed the highest scoring team is retained. This process is repeated for each of the combinations of sensor and control schemes.

To evaluate the effectiveness of the evolved team tactics and the scripted tactics, the best evolved teams for each configuration and the scripted agents are run through the search task a final time. The final search task is also run for 300 seconds, and the *DecayedCoverage* score of each configuration and the scripted agents are returned. The evaluations are only run once for each scenario as the simulation is deterministic. This means that with the same search area configuration and starting point the teams will always perform the same search patterns, and achieve the same score. The evaluation results are presented in the next section.

# 4 Results

In this section we will detail the results of our experiments with learning search tactics. Table 1 shows the *DecayedCoverage* scores for several combinations of inputs/outputs for the MA-HyperNEAT algorithm and the scripted agents, while Figs. 7 and 8 show the learned and scripted search patterns respectively. The first thing to note is that the absolute best performer is the team using the survey area radar and relative velocity and heading outputs. This combination achieved a high score of 10.67 out of a maximum of about 12. The maximum value of 12 is estimated based on the combination of the sensor area of each agent, the number of agents, and the maximum allowed speed. The Radar/VelHeading combination provided the smoothest behavior and tended to develop a concentric circle pattern as can be seen in Fig. 7(b). One agent patrolled along the outer perimeter while the other agents did smaller patrols inside each others tracks. The agents tend to stay in the largest search area and do not venture into the second search area. This result is not surprising since the relative sensors and outputs have produced good results in other problem types.

The tactical combinations varied in their performance depending on the type of output used. The Grid in combination with the VelHeading performed reasonably well with a score of 7.54 beating both the scripted ladder and spiral tactics. This combination lead to a pattern of two agents performing sliding spirals while the third agent drove straight through at seemingly random angles (Fig. 7(a)). When the Grid and Radar inputs were combined this lead to the second best learned tactics with a score of 8.16. This shows that the addition of the radar was able to improve on the base score of the grid, but not enough to bring it up to the level of the radar alone. The resulting pattern produced by this combination resembles the patterns produced by the Grid and Radar separately.

**Table 1.** Results for the search task.

| Learned Search Tactics | | |
|---|---|---|
| Inputs | Outputs | Score |
| Grid | VelHeading | 7.54 |
| GridHeading | VelHeading | 7.06 |
| GridHeading | GridOut | 3.48 |
| GridHeading | AbsOut | 3.54 |
| Grid | VectorOut | 6.18 |
| Radar | VelHeading | 10.67 |
| RadarGrid | VelHeading | 8.16 |
| Scripted Search Tactics | | |
| Spiral | | 6.52 |
| Ladder | | 6.79 |
| Random | | 8.29 |

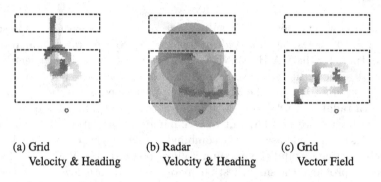

(a) Grid
Velocity & Heading

(b) Radar
Velocity & Heading

(c) Grid
Vector Field

**Fig. 7.** Evolved search tactics

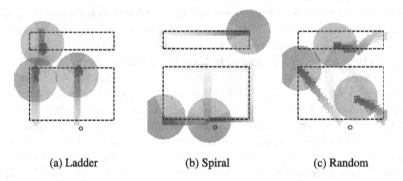

(a) Ladder

(b) Spiral

(c) Random

**Fig. 8.** Scripted search tactics

The concentric circles return but there more exploration outside the larger search area. This lessens the score as the agents loose points when they leave the search area, However this is a more interesting behavior since it starts to cover both areas. The *DecayedCoverage* measure does not take this into account.

The Grid/VectorOut combination scored similarly to the scripted agents with a score of 6.18, meaning it did not perform as well as the other learned tactics. However, The behavior exhibited by this combination is the most interesting. The pattern combines a spiral and a ladder search. Each agent has a larger spiral path but along each run of the spiral it zig-zags in a short ladder search. This can be seen in Fig. 7(c). This combination also behaved better in terms of staying in the search areas more consistently. The primary reason for its lower score comes from the sensor overlap during the zig-zag periods.

The Grid in combination with the GridOut or AbsOut output techniques were the worst performers. These combinations tended to lead to agents that sped straight across the map and embedded themselves against the wall. These techniques will be abandoned during our future work.

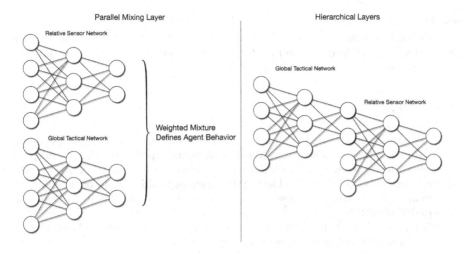

**Fig. 9.** Layered learning will allow the tactical and local networks to work together.

Overall these results show that MA-HyperNEAT is able to learn team search tactics that perform as well or in some cases better than hand coded canonical search tactics. The relative sensor and control schemes produce the best coverage scores, while the tactical sensors and controls tend to explore more of the area. In future experiments the fitness function will be adjusted to provide better feedback for tactics that cover multiple search areas. The combination of the tactical and relative sensors improved the fitness for the tactical sensor, but not enough to overcome the difference. It would be interesting to see how the combination would perform in a layered approach with a tactical network and a local network working together rather than a single network trying to incorporate both.

## 5   Conclusion

As we move into the future where drones and unmanned systems outnumber out human operators it is important that we develop strong and robust autonomous systems to allow each operator to control a larger number of agents. In this work we have explored the use of the MA-HyperNEAT algorithm for the development of team tactics in the area search task. Our results show that the MA-HyperNEAT algorithm is able to develop search tactics that perform as well as, and in some cases outperform, hand coded solutions while providing novel patterns. The results leave us room for improvement and will inform out future work on the subject. While the tactical sensors and controls didn't provide the strongest performance they did lead to interesting behaviors. Our next experiments will try to combine the tactical and local sensors by using layered learning of separate networks that work together rather than trying to combine them into a single network. Figure 9 shows two methods for combining tactical and local

networks. This work will also be extended by incorporating elements of interactive evolution thus allowing a human user to determine which patterns they
prefer. This should lead to novel patterns that are more desirable from the user
standpoint as well as more efficient.

# References

1. Aaltonen, T., Adelman, J., Akimoto, T., Albrow, M.G., González, B.Á., Amerio,
   S., Amidei, D., Anastassov, A.: Measurement of the top-quark mass with dilepton
   events selected using neuroevolution at CDF. Phys. Rev. Lett. **102**, 152001 (2009)
2. Cardamone, L., Loiacono, D., Lanzi, P.L.: Learning to drive in the open racing car
   simulator using online neuroevolution. IEEE Trans. Comput. Intell. AI Games **2**,
   176–190 (2010)
3. D'Ambrosio, D.B., Lehman, J., Risi, S.: MA-HyperNEAT AgentSimulator. http://
   eplex.cs.ucf.edu/software/AgentSimulator_v1_0.zip
4. D'Ambrosio, D.B., Lehman, J., Risi, S., Stanley, K.O.: Evolving policy geometry
   for scalable multiagent learning. In: Proceedings of the Ninth International Conference on Autonomous Agents and Multiagent Systems (AAMAS 2010), pp. 731–738
   (2010)
5. D'Ambrosio, D.B., Stanley, K.O.: Generative encoding for multiagent learning. In:
   Proceedings of the 10th Annual Conference on Genetic and Evolutionary Computation GECCO 2008, vol. 2008, p. 819 (2008)
6. Dragoni, A.F., Giorgini, P.: Belief revision through the belief-function formalism
   in a multi-agent environment. In: Müller, J.P., Wooldridge, M.J., Jennings, N.R.
   (eds.) Intelligent Agents III Agent Theories, Architectures, and Languages. LNCS,
   vol. 1193, pp. 103–115. Springer, Heidelberg (1997)
7. D'Ambrosio, D.B., Stanley, K.O.: Scalable multiagent learning through indirect
   encoding of policy geometry. Evol. Intel. **6**(1), 1–26 (2013)
8. Friedrich, H., Rogalla, O., Dillmann, R.: Communication and propagation of action
   knowledge in multi-agent systems. Robot. Auton. Syst. **29**(1), 41–50 (1999)
9. Garland, A., Alterman, R.: Preparation of multi-agent knowledge for reuse. In:
   Proceedings of the Fall Symposium on Adaptation of Knowledge for Reuse, vol.
   26, p. 33 (1995)
10. Gauci, J., Stanley, K.O.: Autonomous evolution of topographic regularities in arti-
    ficial neural networks. Neural Comput. **22**, 1860–1898 (2010)
11. Gauci, J., Stanley, K.: A case study on the critical role of geometric regularity in
    machine learning. In: AAAI, pp. 628–633 (2008)
12. Le Pape, C.: A combination of centralized and distributed methods for multi-
    agent planning and scheduling. In: Proceedings of the 1990 IEEE International
    Conference on Robotics and Automation, pp. 488–493. IEEE (1990)
13. Panait, L., Luke, S.: Cooperative multi-agent learning: the state of the art. Auton.
    Agent. Multi Agent Syst. **11**(3), 387–434 (2005)
14. Pugh, J.K., Goodell, S., Stanley, K.O.: Directional communication in evolved mul-
    tiagent teams. Technical report, University of Central Florida, Orlando, FL (2013)
15. Sen, S., Weiss, G.: Learning in multiagent systems. In: Weiss, G. (ed.) Multiagent
    Systems: A Modern Approach to Distributed Artificial Intelligence, Chap. 6, pp.
    259–298. The MIT Press, Cambridge (1999)
16. Stanley, K.O.: Compositional pattern producing networks: a novel abstraction of
    development. Genet. Program Evolvable Mach. **8**, 131–162 (2007)

17. Stanley, K.O., Bryant, B.D., Miikkulainen, R.: Real-time neuroevolution in the NERO video game. IEEE Trans. Evol. Comput. **9**, 653–668 (2005)
18. Stanley, K.O., D'Ambrosio, D.B., Gauci, J.: A hypercube-based encoding for evolving large-scale neural networks. Artif. Life **15**, 185–212 (2009)
19. Stanley, K.O., Miikkulainen, R.: A taxonomy for artificial embryogeny. Artif. Life **9**, 93–130 (2003)
20. Stanley, K.O., Miikkulainen, R.: Competitive coevolution through evolutionary complexification (2004)
21. Stone, P., Veloso, M.: Multiagent Systems: A Survey from a Machine Learning Perspective (2000)
22. Taylor, M.E., Whiteson, S., Stone, P.: Comparing evolutionary and temporal difference methods in a reinforcement learning domain. In: Proceedings of the 8th Annual Conference on Genetic and Evolutionary Computation GECCO 2006, p. 1321 (2006)
23. Turing, A.M.: The chemical basis of morphogenesis. Bull. Math. Biol. **52**, 153–197 (1990)
24. Verbancsics, P., Stanley, K.O.: Evolving static representations for task transfer. J. Mach. Learn. Res. **11**, 1737–1769 (2010)
25. Whiteson, S., Stone, P.: Evolutionary function approximation for reinforcement learning. J. Mach. Learn. Res. **7**, 877–917 (2006)
26. Whiteson, S., Whiteson, D.: Machine learning for event selection in high energy physics. Eng. Appl. Artif. Intell. **22**, 1203–1217 (2009)
27. Yao, X.: Evolving artificial neural networks (1999)

# Patterning and Rhythm Generation

# miRNA Regulation of Human Embryonic Stem Cell Differentiation

Gary B. Fogel[1], Tina Tallon[2], Augusta S. Wong[2],
Ana D. Lopez[2], and Charles C. King[2(✉)]

[1] Natural Selection Inc., San Diego, CA, USA
[2] Pediatric Diabetes Research Center,
University of California, San Diego, La Jolla, CA 92121, USA
chking@ucsd.edu

**Abstract.** Elucidating the role that microRNAs (miRNAs) and signaling transduction play in the directed differentiation of human embryonic stem cells (hESCs) into glucose-responsive, insulin-producing endocrine cells is critical to our understanding of systems biology and the development of cell-based therapeutics. To accomplish this, a biochemical understanding the underpinnings of hESC differentiation bias – the propensity of hESCs to differentiate into cells of a specific lineage – must be described in molecular detail. An inherent aspect of hESC culture is stress, and we hypothesize that stress is largely responsible for differentiation bias. Our results indicate that manipulating stress increases apoptosis and disrupts differentiation. Cells subjected to stress fail to become endocrine precursor cells and retain many characteristics of pluripotent cells. Many stresses induce massive apoptosis and result in a loss of up to 80 % of the cells. A consequence of the reduction in cell density is elevated stress signaling, dramatic changes in cell proliferation, maintenance of pluripotency markers, and a complete absence of transcription factors associated with pancreatic endocrine cell production. Coincident with changes in stress, we observed dramatic changes in correlated miRNAexpression, suggesting that cell stress may modulate miRNA transcription and ultimately hESC differentiation.

## 1 Introduction

### 1.1 miRNA Dynamics and During hESC Differentiation

Small, non-coding, regulatory RNA molecules such have microRNAs (miRNAs) have emerged as key rheostats involved in diverse cellular processes and functions (Bagga et al. 2005; Bartel 2004). MiRNAs regulate post-transcriptional gene networks and function in a manner analogous to transcription factors. Mature miRNAs are partially complementary to one or more messenger RNAs (mRNA), and function typically to downregulate gene expression (Hinton et al. 2012; Bagga and Pasquinelli 2006). There are currently several hundred experimentally validated miRNA genes in the human genome, predicted to target many thousands of mRNAs (Ritchie et al. 2013). The importance of miRNAs in hESC differentiation has been demonstrated in studies where critical components of the miRNA machinery have been knocked out (Hinton et al. 2012; Wang et al. 2007; Suh et al. 2004). In two seminal studies, hESC failed to

© Springer International Publishing Switzerland 2015
M. Lones et al. (Eds.): IPCAT 2015, LNCS 9303, pp. 93–102, 2015.
DOI: 10.1007/978-3-319-23108-2_8

differentiate or endocrine tissue failed to form when miRNA processing was disrupted (Suh et al. 2004; Lynn et al. 2007).

Previously, we used microarrays and next-generation sequencing (NGS) to demonstrate that miRNA expression profiles could distinguish pluripotent hESCs from definitive endoderm (DE), the first step toward pancreas lineage specification Hinton et al. 2010, 2014). In the 2010 paper, microarray analysis of a panel of known miRNAs was performed to profile changes in expression during the transition from pluripotency to DE. Our results identified a unique miRNA signature that characterized early pancreas differentiation at the DE stage. Seventeen unique miRNAs were up-regulated in DE compared to pluripotent cells, suggesting a role for these miRNAs in the first step of commitment to endoderm-derived cell lineages. Notably, we detected robust and lineage-specific expression of miR-375 in hESCs differentiated to DE, implicating it in endoderm formation in addition to its established role in regulating islet cell development and function. These studies were amongst the first that demonstrated that miRNAs drive hESC differentiation. A limitation of this work was the inability to detect novel miRNAs and a large numbers of known miRNAs that were not yet available for chip-based analysis. Therefore, this work was extended using next generation sequencing for both cell populations. Millions of sequencing reads from pluripotent and DE cells allowed us to expand the number of miRNAs that significantly change during DE formation from 17 to 77 miRNAs. The changes were significant, four of the five most highly upregulated miRNAs were previously undetected in DE.

As individual miRNAs were interrogated for specific roles in differentiation, questions about global miRNA dynamics affecting differentiation capacity arose. To address this, we undertook NGS analysis of total miRNA expression at 24-h intervals during the first 10 days of hESC differentiation towards pancreatic endoderm (day 0–day 10). Our biological and computational analysis revealed rapid, daily changes in the expression of specific miRNAs grouped within 4 sub-clusters, coupled with longer patterns of coordinated miRNA changes as cell lineage specificity emerged (Fogeland King, in revision). Alterations of growth factor composition of the media strongly correlated with increases in stress and dramatic changes in miRNA expression. While these data provided essential insights into how the dynamics of miRNA expression regulates hESC differentiation through DE towards pancreatic endoderm, little was known about whether expression of miRNAs during this process was correlated or about how all miRNAs functioned as a network.

## 1.2 Chemical Regulators of Pluripotency

During DE formation, a small number of hESC remain pluripotent, resulting in teratoma formation following transplant. From our signal transduction work, we identified a compound that regulates hESC pluripotency: Gö6976, a protein kinase C inhibitor that enhances exit from pluripotency. The effect of Gö6976 on exit from pluripotency appears to be pleiotropic. Although inhibition of PKC in cells occurs at low concentrations (250 nM), other off target effects that alter signaling also occur during hESC differentiation to DE. One is on signaling through the JAK/Stat pathway (Reyes-Cava and King, in preparation). Given the critical role of miRNAs in the transition of pluripotent hESCs to DE, we wanted to examine whether Gö6976 also disrupted the dynamics of miRNA expression.

## 2 Methods

### 2.1 Stem Cell Culture and Differentiation

CyT49 cells (provided by ViaCyte, San Diego, CA) were maintained on reduced growth factor BD Matrigel at 37 °C, 5 % CO2 in DMEM/F12 supplemented with 20 % knockout serum replacement, glutamax, nonessential amino acids, β-mercaptoethanol, penicillin/streptomycin (Life Technologies, Carlsbad, CA), 4 ng/mL basic fibroblast growth factor (FGF; Peprotech, Rocky Hill, NJ) and 10 ng/mL activin A (R&D Systems, Minneapolis, MN). Differentiation to pancreatic precursor stage was carried out as previously described (Hinton et al. 2014). Details for hESC differentiation are as follows: Day 0 – Activin A (100 ng/ml; R&D Systems, Minneapolis, MN) and Wnt3A (25 ng/ml, R&D Systems) with 0.2 % FBS (Hyclone, Thermo Fisher Scientific, Waltham, MA) in RPMI-1640 (Invitrogen, Carlsbad, CA); days 1 and 2– Activin A with 0.5 % FBS in RPMI-1640; day 3 – Activin A with 2 % FBS in RPMI-1640.days 4–6 – kerotinocyte growth factor (KGF; 50 ng/ml) in RPMI-1640 with 2 % FBS; days 7–8 – KAAD-cyclopamine (0.25 µM, EMD-Millipore, San Diego, CA), retinoic acid (2 µM, Sigma, St. Louis, MO), noggin (50 ng/ml, R&D Systems), 1 % B-27 (Invitrogen) in DMEM (Invitrogen); days 9-10–1 % B-27 (Invitrogen) in DMEM (Invitrogen). Gö6976 (Calbiochem, Danver, MA) was added to the media at days 1, 2, and 3 at a final concentration of 250 nM.

### 2.2 RNA Preparation

To ensure representative samples were generated for each time point during hESC differentiation, the following biological replicates were generated starting from the same passage of CyT49 cells: Day 0 ($n = 8$); Day 1 ($n = 4$); Day 2 ($n = 4$); Day 3 ($n = 4$); Day 4 ($n = 6$). For each 24-h interval of differentiation, samples were prepared as follows. Cells were lysed in Trizol and RNA was extracted by the manufacturer's recommended protocol (Life Technologies, Carlsbad, CA). Resultant RNA was treated with Turbo DNAse (Life Technologies) for 30 min. DNAse-treated RNA was purified by sequential extraction in acid phenol:chloroform (5:1), followed by chloroform alone, then precipitated in 4 volumes ethanol. Small RNA libraries were prepared using the Small RNA 1.0 Sample Preparation Kit (Illumina, Inc., San Diego, CA). A band of RNA ranging from 18–30 nt was cut from a 15 % TBE-urea gel and RNA was extracted according to the manufacturer's recommended protocol. After ligation of 5' and 3' adaptors, bands of 40–60 and 70–90 nt, respectively, were cut from the gel and RNA was again extracted as described above, followed by RT-PCR amplification. Finally, a ~ 92 bp band of small RNA library was purified from the gel. The library was validated on an Agilent 2100 Bioanalyzer using the DNA100 chip and quantified using a Roche LightCycler 480. Ten picomoles were run per flow cell in an

Illumina GAII sequencer using a v4 Cluster generation kit and a v5 sequencing kit for 36 cycles with Illumina Sequencing Primer Read 1 Mix.

## 2.3   Network Analysis

Given statistical correlation measurements for all miRNA pairs, network graphs can be constructed where vertices represent miRNAs and edges connect miRNAs that have a strong positive or negative correlation (that exceeds some threshold). Remondini et al. 2007, analyzed such network graphs that represent strongly correlated expression time series for microarray gene expression data. In particular, they considered vertices that have a high ratio of betweenness centrality, $b$, to connectivity degree, $k$. The betweenness centrality of a vertex V is the number of shortest-paths between pairs of vertices U and W (other than V) that traverse V. The connectivity degree of a vertex V is the number of edges incident on the vertex. In the case of miRNA expression, vertices with a high value of $b/k$ may represent miRNAs operating early in an expression cascade.

# 3   Results

## 3.1   Correlated Expression of miRNAs During hESC Differentiation

After measuring pairwise correlations between the 694 miRNAs that were expressed over all 10 days, we obtained the distribution of correlation coefficients. Generating artificial time series from the distribution of log expression values in our data, 20,000 random time series were obtained. We determined that absolute correlation values above 0.98 and 0.99 corresponded to α-values of 0.05 and 0.01, respectively, thus establishing useful significance levels for our real data. Selected miRNAs associated with pluripotency/differentiation and development/growth dominated the list of the top 20 miRNAs that change during early differentiation (Table 1). Most of the changes were correlated within specific miRNA clusters. For example, expression of the miRNAs in the miR-371/372/373 cluster were highly correlated, as were miR-302a, miR-302c, and miR-302d (data not shown). Interestingly, miR-302b and miR-367 expression profiles did not correlate strongly with other miRNAs in their cluster, suggesting that expression even within a specific miRNA cluster can be differentially regulated. With the exception of miR-103a-3p, the expression profiles of selected miRNAs associated with pluripotency/differentiation or development/growth did not correlate strongly with those of other miRNAs. The miR-103a-3p expression profile correlated strongly with those of 13 different miRNAs (miR-30e-5p, miR-148b-3p, miR-151a-5p, miR-151a-3p, miR-181c-5p, miR-181d, miR-197-3p, miR-301a-3p, miR-338-3p, miR-342-3p, miR-429, miR-574-3p, and miR-769-5p) (Fig. 1). The profile of expression follows the pattern of low, but sustained expression during DE formation, followed by a rapid increase in miRNA expression during cell expansion, another plateau as cells adjust to a change in media containing retinoic acid, cyclopanine, and noggin, and finally a second burst of miRNA expression as these growth factors are removed.

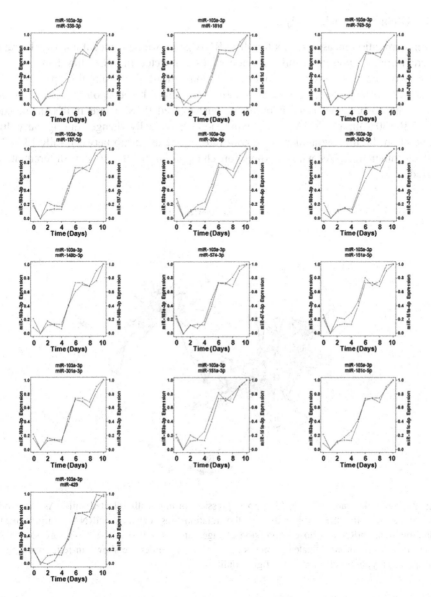

**Fig. 1.** Correlation of miR-103a-3p expression profile with selected miRNAs. Thirteen different non-clustered miRNAs have coordinated expression with miR-103a during the 11 day differentiation from pluripotency to pancreatic precursor cell. Each miRNA has an absolute correlation value above 0.99 which corresponds to an α-value of 0.01. This is one example of a single pattern of correlated miRNA expression during differentiation. Many others were observed (data not shown).

## 3.2   Global Network Analysis

Given correlation measurements for all miRNA pairs, we constructed a network graph where vertices represented miRNAs and edges connected miRNAs that had a strong positive or negative correlation (that exceeds some threshold). Using this approach, we found six miRNAs to have relatively large *b*/*k* values, based upon a network graph where the absolute value of correlation must exceed 0.98. These miRNAs include hsa-miR-130a-3p; a miRNA whose expression dramatically changes during early differentiation. Figure 2 graphically demonstrates the relationship between miR-130a-3p and the other miRNAs whose expression changes during the 10 day differentiation period.

mir: 130a–3p clst=2 b=132 k=2 r=66

**Fig. 2.** Network graph of miR-130a-3p expression profile with selected miRNAs. Network analysis allows for the exploration of the relationships between miRNAs during hESC differentiation. miRNAs whose expression changes are related to miR-130a-3p are shown in black, then levels of coordinated expression in decreasing order are shown in red, green, blue, teal, pink, grey, and yellow (Colour figure online).

## 3.3   Effect of Gö6976 on miRNA Expression During DE Formation

Cells treated with Gö6976 for _only_ days one, two, and three of the ten day differentiation protocol lost all pluripotency markers, but did not express markers of pancreatic endoderm. Results were confirmed by both Western blotting and immunofluorescence (data not shown). Additionally, PCR arrays and Western blotting found that stress signaling pathways and gene expression were elevated further in Gö6976-treated hESCs. Gö6976-treated hESCs undergo massive apoptosis, losing up to 70 % of the plated cells, compared with the ∼10 % of cells lost to apoptosis in untreated cells. Within 5 days, proliferation increases and the cells recover. However, after recovery, Gö6976-treated cells appear dramatically different from untreated cells under the microscope. NGS sequencing of miRNAs from Gö6976-treated cells after 4 days revealed a dramatic change in miRNA expression. Table 1 shows the top 20 miRNAs expressed in RPM (reads per million) in hESC after differentiation to DE. Consistent with our previously published work, a number of miRNAs we have identified as critical for DE formation are present. Upon treatment with Gö6976, expression of these miRNAs drop (column 1 vs. column 2). This is not due to a general blocking of miRNA expression by treatment with Gö6976, because other miRNAs increased significantly (column 3 vs. column 4). The decrease in expression of various miRNAs was confirmed by TaqMan analysis (not shown). Decreased hESC plating density at the initiation of differentiation recapitulates the effect of Gö6976 treatment.

**Table 1.** Changes in miRNA expression in the presence of Gö6976. The miRNAs listed on the left were upregulated during DE formation in CyT49 cells at Day 4 (Av. RPM φ; shaded green). Upon addition of Gö6976, expression of these miRNAs drops (Av. RPM + Gö6976; shaded red). The miRNAs listed on the right were upregulated in cells treated with Gö6976. (The last column on the right shows expression in untreated cells.)

| miRNA | Av. RPM φ | Av. RPM + Gö6976 | miRNA | Av. RPM φ | Av. RPM + Gö6976 |
|---|---|---|---|---|---|
| hsa-miR-375 | 106087.3 | 271.70 | has-miR-939-5p | 0.08 | 21578.9 |
| hsa-miR-372 | 74219.5 | 0.00 | hsa-miR-491-3p | 0.70 | 19841.1 |
| hsa-miR-103a-3p | 47417.1 | 0.00 | hsa-miR-3661 | 1.11 | 19384.2 |
| hsa-miR-371a-5p | 46690.9 | 0.00 | hsa-miR-3663-5p | 0.00 | 17182.6 |
| hsa-miR-92a-3p | 23448.3 | 0.00 | hsa-miR-1275 | 9.00 | 7262.7 |
| hsa-miR-21-5p | 15076.9 | 0.00 | hsa-miR-519b-5p | 0.77 | 5897.7 |
| hsa-miR-320a | 12410.2 | 0.00 | hsa-miR-1228-3p | 0.06 | 5064.3 |
| hsa-miR-25-3p | 10584.4 | 0.00 | hsa-miR-105-3p | 0.22 | 5059.6 |
| hsa-miR-191-5p | 9453.1 | 91.30 | hsa-miR-1302 | 0.00 | 4977.6 |
| hsa-miR-423-5p | 8207.5 | 98.70 | hsa-miR-25-3p | 1507.28 | 3175.8 |
| hsa-miR-302b-3p | 6901.7 | 0.00 | hsa-miR-671-3p | 22.31 | 2835.1 |
| hsa-miR-302a-3p | 6498.1 | 0.00 | hsa-miR-876-3p | 151.92 | 10942.5 |
| hsa-miR-140-3p | 5978.5 | 0.00 | hsa-miR-518e-5p | 0.07 | 2494.4 |
| hsa-miR-302d-3p | 5480.7 | 0.00 | hsa-miR-1287 | 0.77 | 2200.3 |
| hsa-miR-340-5p | 5397.4 | 89.10 | hsa-miR-1305 | 6.01 | 1781.7 |
| hsa-miR-378a-3p | 5027.3 | 0.00 | hsa-miR-3675-3p | 7.06 | 1579.0 |
| hsa-miR-302a-5p | 4917.2 | 0.00 | hsa-miR-3685 | 0.00 | 1550.1 |
| hsa-miR-130a-3p | 4444.6 | 96.20 | hsa-miR-29c-5p | 0.00 | 1545.7 |
| hsa-miR-373-3p | 4242.4 | 0.00 | hsa-miR-3679-5p | 0.47 | 1510.0 |
| hsa-miR-200c-3p | 4045.2 | 31.20 | hsa-miR-3146 | 0.18 | 1496.6 |

We next wanted to determine whether changes in cellular architecture observed upon Gö6976 treatment were a result of the drug or the apoptosis. Results from RT-PCR experiments found that decreased cell plating density yielded results identical to Gö6976 treatment (Fig. 3). Cells lost the ability to become pancreatic lineage (no expression of PDX1 and FoxA2), while expression of the mesendoderm marker Brachyury increased, indicating cells shift into non-directed differentiation.

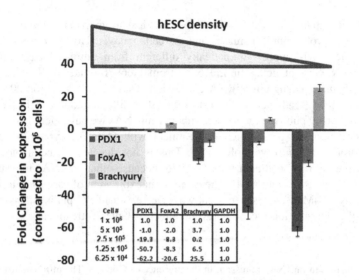

**Fig. 3.** Plating hESC at decreasing densities alters attenuates expression of pancreatic markers and increases expression of the mesendoderm marker Brachyury. Cyt49 cells from the same passage were plated at decreasing densities. As density was decreased, the cell stress increased, and expression of pancreatic markers PDX1 and FoxA2 decreased. During these differentiations, expression of Brachyury, a mesendoderm marker increased significantly.

## 4 Discussion

Overall, correlated miRNA expression profile analysis revealed two distinct and unexpected trends. Firstly, not all mature miRNAs within a miRNA cluster originating from the same locus and likely subjected to the same transcriptional regulation followed the same expression profile. Of the many clusters examined, including the miR-371/372/373 and the miR-520 cluster, many had intra-cluster correlated expression profiles. However, some clusters such as the miR-302/367 cluster had selected miRNAs whose expression profiles were not correlated and varied 100-fold between specific miRNAs (data not shown). This suggests differential regulation of expression, processing, and/or degradation. Secondly, multiple seemingly unrelated miRNAs have nearly identical expression profiles. This is exemplified by miR-103a, whose expression correlates strongly with 13 other miRNAs (Fig. 1). While there is a possibility that some of the expression profiles match by chance, it is unlikely that all do. While miR-103a was the only miRNA within the limited selection of pluripotency/differentiation and development/growth cohort we examined in detail to correlate strongly with other miRNA, it was far from the only one. Additionally, we observed anti-correlation of selected miRNA expression, suggesting that under certain circumstances, miRNAs might suppress expression of other miRNAs (data not shown). Coordinated regulation of miRNA expression has not been well documented, but our results indicate that this might be a mechanism by which expression of multiple proteins required to shift cell fate can be simultaneously down-regulated. Network analysis of miRNAs (Fig. 2) supports this hypothesis.

In an attempt to explore how signal transduction regulates cell fate decisions, the protein kinase C inhibitor Gö6976 was incubated with hESC during DE formation. The net result was decreased expression of pluripotency markers coupled with increased apoptosis. After Gö6976 treatment, no pancreatic precursor cells were formed as measures by expression of FoxA2 and PDX1. Exploring the early stages of differentiation, we found that plating cells at lower density also prevented formation of pancreatic precursors, suggesting that the massive apoptosis associated with Gö6976 treatment helped direct cell fate. Both lower density plating and Gö6976 treatment were found to increase cell stress as measured by increased expression of stress-related genes and kinases. In addition, Gö6976 treated cells had dramatically altered expression of miRNAs. We hypothesize that stress related acutely to apoptosis, but ultimately to cell density result in misregualtion of miRNA expression which alters cell fate (Fig. 4). Additional work regarding the nature of stress, the means through which this information is conveyed in the cell, and the resulting cellular development is required.

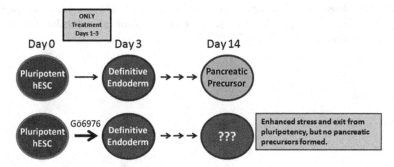

**Fig. 4. Chemical modulation of hESC fate through increased cell stress.** Pancreatic cell fate can be altered by increased stress early (Days 1–3) during differentiation. Treatment of cells with Gö6976 increased expression of markers of cell stress and altered miRNA expression. The net result of this is cells of mixed fate. Whether miRNAs are drivers of this process is currently being explored.

**Acknowledgements.** Funding for these studies was provided by the Larry L. Hillblom Foundation (CCK) and the California Institute for Regenerative Medicine (CIRM to CCK). The authors would also like to thank Sevan Ficici for his assistance.

# References

Bagga, S., et al.: Regulation by let-7 and lin-4 miRNAs results in target mRNA degradation. Cell **122**, 553–563 (2005)

Bartel, D.P.: MicroRNAs: genomics, biogenesis, mechanism, and function. Cell **116**, 281–297 (2004)

Hinton, A., Hunter, S., Reyes, G., Fogel, G.B., King, C.C.: From pluripotency to islets: miRNAs as critical regulators of human cellular differentiation. Adv. Genet. **79**, 1–34 (2012)

Bagga, S., Pasquinelli, A.E.: Identification and analysis of microRNAs. Genet. Eng. (N Y) **27**, 1–20 (2006)

Ritchie, W., Rasko, J.E., Flamant, S.: MicroRNA target prediction and validation. Adv. Exp. Med. Biol. **774**, 39–53 (2013)

Wang, Y., Medvid, R., Melton, C., Jaenisch, R., Blelloch, R.: DGCR8 is essential for microRNA biogenesis and silencing of embryonic stem cell self-renewal. Nat. Genet. **39**, 380–385 (2007)

Suh, M.R., et al.: Human embryonic stem cells express a unique set of microRNAs. Dev. Biol. **270**, 488–498 (2004)

Suh, N., et al.: MicroRNA function is globally suppressed in mouse oocytes and early embryos. Curr. Biol. **20**, 271–277 (2010)

Lynn, F.C., et al.: MicroRNA expression is required for pancreatic islet cell genesis in the mouse. Diabetes **56**, 2938–2945 (2007)

Hinton, A., et al.: A distinct microRNA signature for definitive endoderm derived from human embryonic stem cells. Stem Cells Dev. **19**, 797–807 (2010)

Hinton, A., et al.: sRNA-seq analysis of human embryonic stem cells and definitive endoderm reveal differentially expressed microRNAs and novel isomiRs with distinct targets. Stem Cells **32**, 2360–2372 (2014)

Remondini, D., Neretti, N., Franceschi, C., Tieri, P., Sedivy, J.M., Milanesi, L., Castellani, G.C.: Networks from gene expression time series: characterization of correlation patterns. Int. J. Bifurcat. Chaos **17**(7), 2477–2483 (2007)

# Motifs Within Genetic Regulatory Networks Increase Organization During Pattern Formation

Hamid Mohamadlou, Gregory J. Podgorski, and Nicholas S. Flann$^{(\boxtimes)}$

Department of Computer Science, Department of Biology, Logan, Utah
hamid.mohamadlou@aggiemail.usu.edu,
{gregory.podgorski,nick.flann}@usu.edu

**Abstract.** Motifs are small gene interaction networks that frequently occur within larger genetic regulatory networks (GRNs). However, it is unclear what evolutionary and developmental advantages motifs provide that have led to this enrichment. This study seeks to understand how motifs within developmental GRNs influence the complexity of multicellular patterns that emerge from the dynamics of the regulatory networks. A computational study was performed by creating Boolean intracellular networks with varying inserted motifs within a simulated epithelial field of embryonic cells. Each cell contains the same network and communicates with adjacent cells using contact-mediated signaling. Comparison of random networks to those with motifs demonstrated that: (1) Bistable switches that encode mutual inhibition simplify both the pattern and network dynamics. (2) All other motifs with feedback loops increase information complexity of the multicellular patterns while simplifying the network dynamics. (3) Negative feedback loops affect the dynamics complexity more significantly than positive feedback loops. (4) Feed forward motifs without feedback have little effect on the complexity of patterns formed.

**Keywords:** Network motifs · Kolmogorov complexity · Pattern formation · Genetic regulatory networks

## 1 Introduction

Multicellular organisms contain a large variety of cellular patterns. For instance, Fig. 1 illustrates a *Drosophila melanogaster* embryo in which muscle and nervous system structures interconnect through sensory and activation signaling. These patterns are formed during development and are a consequence of genetic regulatory networks (GRNs) that operate within cells and that respond to communication between cells [1–3]. GRN's are networks of interacting genes where the expression or non-expression of genes determines the expression state of other genes. The dynamics of GRNs determine the gene expression profile for each cell leading to spatial patterns of cellular differentiation. This process is repeated to implement an organism's body plan, and subsequence morphology [4].

© Springer International Publishing Switzerland 2015
M. Lones et al. (Eds.): IPCAT 2015, LNCS 9303, pp. 103–113, 2015.
DOI: 10.1007/978-3-319-23108-2_9

GRNs contain subnetworks of genes referred to as motifs [5,6]. Motifs are detected at a higher frequency than would be expected in random networks. Computational biologists have hypothesized that motifs play a determinative role in cell function [7,8]. However, their influence on pattern formation during development is poorly understood. This work presents a computational study aimed at understanding how the presence of motifs within intracellular networks changes the GRN dynamics and the emergent multicellular patterns.

In particular, this study measures the complexity of the organization of the dynamics and patterns. As can be seen in Fig. 1, patterns can involve complex arrangements of specialized regions and interconnections that develop later, or earlier patterns such as simple segmentation [9] or mosaics arrangements [10,11]. To quantify the level of organization we employ Kolmogorov complexity, also known as algorithmic complexity [12]. Such methods measures the information contained within an object, such as a cellular pattern, by considering the size of the algorithm needed to generate the object, hence the term algorithmic complexity. The smaller the algorithm, the simpler the pattern. The calculations of Kolmogorov complexity is detailed in the methods section following.

To understand how GRNs regulate biological events, scientists have developed mathematical and computational models to generate predictions and explain experimental observations. Among these modeling approaches is to represent a GRN as a Boolean network in which the activity of a gene is either on or off, determined by a set of logical functions over the activity of other genes [13]. It is this modeling framework that we apply in this study.

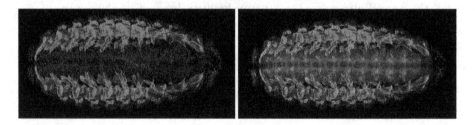

**Fig. 1.** Ventral view of stage 16 *Drosophila melanogaster* embryo immunostained for tropomyosin (green; a protein expressed in muscle), Pax 3/7 (blue; a regulatory protein expressed in central nervous system nuclei and ectoderm), and HRP (red; neurons). All nuclei shown in gray (DAPI). Courtesy of Julieta María Acevedo and Lucas Leclere, Marine Biological Laboratory, Woods Hole, www.mbl.edu/dev.biologists.org/(Color figure online).

To evaluate the influence of motifs on network dynamics and patterns, we design GRNs that are embedded into cells arranged in a 2D grid, simulating an epithelium. Such abstractions of the epithelium have been employed successfully in many developmental systems, such as the cellularized *Drosophila* embryo [9], and the sensory epithelia of the developing vertebrate retina [14] and the inner ear [10]. Each cell contains an identical Boolean network, referred to as a complete network. We explore the influence of the best-understood motifs on network dynamics and multicellular patterns by inserting them into randomly generated Boolean networks.

## 2 Motifs Within Gene Regulatory Networks

It is believed that motifs increase the modularity of gene regulatory networks by performing relatively independent tasks [15,16] such as decision making, signal processing and communication. The modular organization of biological structure is supported by experimental studies from pathogen structure, gene networks, and protein-protein interaction networks [17]. For example, Kim et al. [15] studied the connected subset of protein networks in protein-protein interaction data for budding yeast. Their analysis suggests that the yeast protein network is significantly modular, and it contains various motifs.

Motifs are defined as a set of interconnected genes that produce a distinct function, regardless of whether they are structurally isolated within a network. Motifs frequently occur and consist of few interacting genes [8]. Motifs were first noted in *Escherichia coli*, where they were detected at a higher frequency than would be expected in random networks. Since then multiple motifs have been identified in bacteria and yeast [20], the immune system [21], and *Drosophila* [15]. This finding

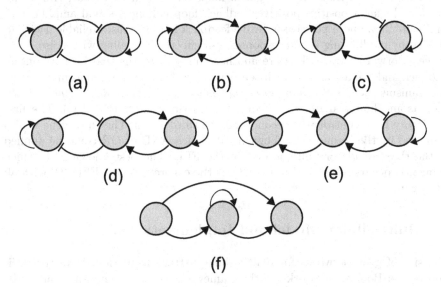

**Fig. 2.** (a) A positive feedback loop (a double inhibitory loop with two positive autoregulatory loops). (b) A positive feedback loop (a double excitatory loop with two positive autoregulatory loops). (c) A negative feedback loop [18] with two positive autoregulatory loops. (d) Coupled positive-positive feedback loops. (e) Coupled positive-negative feedback loops. (f) The type-1 coherent feed-forward loop [19].

suggests that motifs are building blocks of transcription networks and that they may have evolved to achieve specific regulatory behaviours in cellular transcription networks [18]. Regulatory motifs have be found in regulatory networks that perform two distinct functions: (1) Developmental networks that guide differentiation and cell fate determination by transducing signals into irreversible cell-fate decisions [22, 23]; and (2) Sensory networks that respond to signals such as stresses and nutrients rapidly and make reversible decisions [5].

The motifs that are associated with developmental networks are commonly comprised of feedback loops. Positive-feedback loops are most common and are made up of two transcription factors that regulate each other. There are two kinds of positive-feedback loops, a double excitatory loop (Fig. 2(b)) and a double-inhibitory loop (Fig. 2(a)). The regulatory dynamics of these gene pairs coupled by positive feedback loops often results in two or more steady states and is referred to as multistability [18]. Positive feedback loops amplify signals and elongate the time required to reach a steady state, referred to as a network attractor [20]. This slowed response can be helpful when a cell makes significant decisions such as irreversible cell specification and apoptosis. Unlike positive feedback loops, negative feedback loops (Fig. 2(c)) often enhance attractor stability. They also function as noise filters and make cells more robust to signal noises. Also, positive and negative feedback loops are coupled into structures containing two feedback loops, such as positive-positive, positive-negative and negative-negative feedback loops (Fig. 2(d and e)). Coupled feedback loops perform functions that single feedback loops cannot. In particular, Kim et al. [18] found that a positive-positive feedback loop enhances signal amplification and bistability, and a positive-negative feedback loop increases reliable decision-making by modulating signal responses and effectively dealing with noise.

Feed-forward loops (FFL) are another family of motifs that do not contain feedback and are associated with sensory networks. FFL are found in a variety of organisms such as *Saccharomyces cerevisiae*, *Bacillus subtilis*, *Caenorhabditis Elegans* and humans [19]. FFLs consists of a three genes (Fig. 2(f)). The first regulatory gene controls the second and the third genes. The third gene is also regulated by the second gene. Logical gates such AND or OR could be applied to the three regulatory interactions in the FFL. The best known FFL which frequently occurs in (*E. coli* and yeast), is the coherent type-1 FFL [24] with all AND gates.

## 3    Multicellular Model and Implementation

The size of gene networks for multicellular pattern formation drove the decision to use Boolean networks as the framework for this computational study. A Boolean network is a simplified model of a genetic regulatory network. In this application, each gene is represented as a network node that takes binary values (1 for expressed and 0 for not expressed). The state of a gene (0 or 1) is determined by its Boolean function defined as the expressions of AND, OR, NOT on the inputs from other genes. These inputs are represented as directed edges in

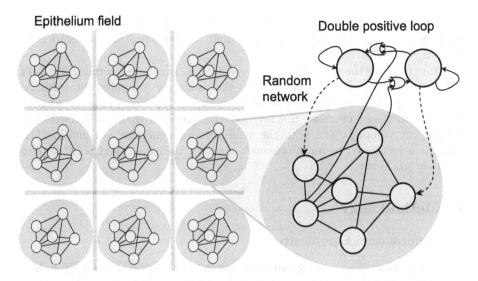

**Fig. 3. Motif insertion.** Example motif insertion into a random GRN. Dashed arrows represent random outgoing signals from the motif. Outgoing and incoming signals from and to the random GRN are randomly connected to genes within the random network.

the network graph. Boolean networks provide a qualitative description of gene states and their interactions, first introduced by Kauffman [13,25].

This work extends our previous study [3] that studied the role of different dynamical regimes (ordered, critical or chaotic) the complexity of multicellular pattern formation. For this study, all networks were created in the critical domain and then motifs were inserted (Fig. 3). The simulation model was unchanged along with the epithelial model as a lattice of cells, with each cell holding a complete Boolean network. Cell-cell signaling was implemented in the model as an edge connecting the state of one gene in a cell to an input of a Boolean function of one or more of its neighbors. Such genes are called communicating genes. The number of communicating genes is referred to as the signaling bandwidth.

Signaling bandwidth was set to half of the total number of genes in a cell as our previous study showed that this configuration established the most efficient cell-cell signaling. We employed directional signaling between adjacent cells were cells connect to their north-south and/or east-west neighbors. These directions corresponded to the anterior-posterior and dorsal-ventral embryonic axis [3].

The state of each cellular GRN is initialized randomly by setting the state of each gene to 0 or 1. Randomly generated logic functions are assigned to networks as the transition rules used to determine the state of genes [3]. The state of the system during simulation is clocked synchronously until a steady or cyclic state (in up to 300 repeats) is reached for all individual cells. When the state of genes change in a repetitive cycle or reach a fixed state, then cell are in attractor state [13]. Since cells are connected through directional signaling, neighboring cells may converge to different attractors, forming a regular pattern, where cells in the same attractor have differentiated into the same cell types. The state of all the genes as the networks along with multicellular patterns are recorded for analysis of information content.

After running the randomly-generated GRNs, single and coupled feedback and feedforward loops are inserted into the randomly generated intra-cellular GRN Fig. 3). The network is run again with the inserted motifs to identify the attractors and visualize the multicellular patterns that are formed. The Kolmogorov complexity of both the gene network dynamics and multicellular patterns is computed by an information theoretic measure called Set Complexity described below.

# 4   Methodology

## 4.1   Information Complexity

Set Complexity [26] is an information complexity metric derived from Kolmogorov's original work on algorithmic complexity. In this seminal paper by Galis et al., the method was developed explicitly to measure the information contained in sets of biological data at different scales: the molecular, sequence and network. Recently, Set Complexity has been applied to study modularity in biological networks [27] and large-scale genetic interaction networks [28]. In this work, we extend the approach to the analysis of multicellular patterns and the network dynamics that create them.

Kolmogorov complexity of an object is the size of the shortest program that can produce that object. The exact calculation is undecidable and, therefore, cannot be computed as defined. As an approximation, the compression size of an object is used. Kolmogorov analysis defines the similarity of two objects as the size of the shortest program that can translate one object to the other. This measure is approximated by the Normalized Compression Distance (NCD) [29], where the NCD of two objects is 0.0 if they are identical and 1.0 if they are random. Set Complexity combines the analysis of single objects and pairwise objects to arrive at a complexity measure of a set of objects, given in Eq. 1. By employing NCD as a metric to evaluate the similarity of all pairs of objects in a set, set complexity discounts the influence of the pairs of objects that are randomly related or redundant. As long as any object can be encoded as a string, Set Complexity can compute the information content that resides in the set.

Set Complexity of a set of $n$ strings $S = \{s_1, \ldots, s_n\}$ is defined:

$$\Psi(S) = \frac{1}{n(n-1)} \sum_{s_i \in S} C(s_i) \sum_{s_j \neq s_i} NCD(s_i, s_j)(1 - NCD(s_i, s_j)) \qquad (1)$$

where $C(s_i)$ is the compression size of string $s_i$. The term $NCD(s_i, s_j)$ $(1 - NCD(s_i, s_j))$ is maximized when $NCD(s_i, s_j) = 0.5$, which occurs when $C(s_i + s_j) \simeq C(s_i)/2 - C(s_j)$, assuming $C(s_i) > C(s_j)$. The influence of similar and dissimilar strings is discounted by $NCD(s_i, s_j)(1 - NCD(s_i, s_j))$ because

$NCD(s_i, s_j)$ will be near 1.0 or 0.0. In this implementation, $bZip$ is used as a compression algorithm.

A one-to-one mapping is required to encode an object into a string so that no information is lost. The method by which each GRN, its temporal dynamics, and the spatial pattern produced are encoded as a string is described in [3].

## 5   Results

In this section, we explore the effect of insertion of the best-known motifs illustrated in Fig. 2 into randomly generated coupled GRNs. For each motif, 600 random networks were generated and then each network was modified by the insertion of that regulatory motif. To obtain the data for analysis, each network was executed from initial random conditions and the dynamics and resulting pattern recorded. Following this step, the network data is grouped into sets of 10 and the set complexity computed for the networks, the dynamics, and the pattern.

Figure 4 shows the influence of insertion of a positive feedback loop (Fig. 2(b)) into a random GRN. The vertical axis of both graphs is the pattern complexity; on the left shows the relationship with dynamics complexity, on the right shows the relationship with network complexity. Each point on the graph corresponds to a set of 10 runs described above.

Insertion of a double positive feedback loop into random GRNs clearly modifies the behavior of the network. The addition of the motif increases pattern complexity significantly while slightly reducing the complexity of the dynamics (Fig. 4 left). Little effect on network complexity is observed (Fig. 4 right).

Figure 5 illustrates the average network, network dynamics and pattern complexity for 60 network sets for each of the motifs. The influence of each motif is represented as an arrow. The start of the arrow is the set of random networks and is the same for all motifs. The arrowhead is the average complexity of the set of all networks with the indicated motif inserted. The insertion of a

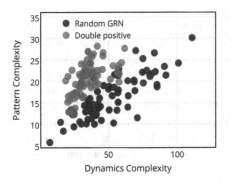
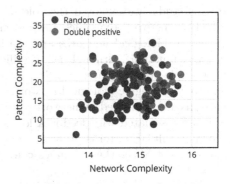

**Fig. 4.** Effect of insertion of a double positive feedback loop on network, network dynamics and pattern complexity.

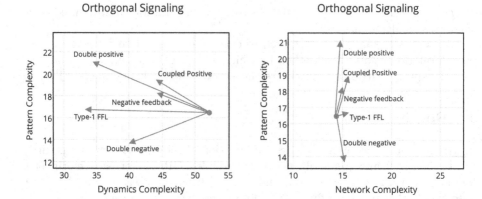

**Fig. 5.** (a) Effects on network dynamics and pattern complexity of inserting regulatory motifs into random GRNs. Average dynamics and pattern complexity for 60 random sets of GRNs (arrow tails) and 60 sets of GRNs with the indicated inserted motifs (arrow heads). (b) Effects on network and pattern complexity of inserting regulatory motifs into random GRNs.

negative feedback loop, a double negative feedback loop or a double positive feedback loop all have the same qualitative effect of decreasing network dynamics complexity and increasing pattern complexity. However, a double positive loop increases pattern complexity much more than either of the negative feedback loops (Fig. 5(a)). Insertion of feed-forward motifs decreases dynamics with almost no effect on the pattern complexity. Only the double negative motif leads to a reduction in pattern complexity. Statistical analysis confirms that insertion of these motifs makes a significant change in the average network dynamics and patterns complexity.

# 6   Discussion

In this study, we explored the role of common regulatory motifs in network structure, network dynamics and pattern complexity. These motifs frequently appear in biological networks and are thought to play critical roles in overall network function. Although the significance of these motifs have been shown in multiple studies, there is a lack of computational analysis to explore how and to what degree biological network dynamics and the resulting multicellular patterns are influenced by network motifs. The results show that network motifs that are associated with feedback loops increase the information complexity of the multicellular patterns. Another important observation was negative feedback loops do not affect the dynamics complexity significantly as positive feedback loops do.

We hypothesise that variation in dynamics complexity associated with adding different types of motifs to random GRNs originates from the time it takes for the GRN to reach a steady state and the proportion of single state versus cyclic attractors produced by the GRN. Since the motifs studied here act as multistable switches, they simplify the complex cyclic attractors to attractors with a few states. Gene expression in feedback loops reach a steady state quickly, and so reduce the length and complexity of cyclic attractors. We hypothesize that this is why in all cases dynamics complexity decreases from that of the original random networks. The rate of dynamics complexity reduction associated with the addition of the motifs with negative feedback loops is significantly lower than for positive loops. Unlike positive feedback loops, negative feedback loops do not increase the time to reach steady states [18]. Therefore, they don't effect the dynamics complexity noticeably. As results show, negative feedback loop motifs (such as single negative feedback loop and coupled positive-negative loops) have the lowest reduction in their dynamics complexity.

All the motifs with feedback loops impact the pattern complexity. In fact, the only motifs in this study that has no effect on pattern complexity are feedforward loops. This observation confirms the association of feedback loop motifs with developmental networks that mediate important cell fate decision. The increase in information complexity of the multicellular patterns shows more predictable cell-fate decisions have been taken. Only double negative motifs decrease the complexity of the patterns formed. We hypothesize that this is due to the bistable behavior of these motifs, where they rapidly converge to one state or another producing few distinct non-cyclic attractors.

Overall, we find that all motifs with feedback decrease the complexity of the dynamics. By decreasing the complexity of the dynamics, these motifs could increase the robustness in pattern formation in the presence of noise as shown in [8]. Feedback motifs tend to increase pattern complexity while decreasing dynamics complexity. This implies that these motifs increase the "efficiency" of pattern formation where simpler dynamics leads to more complex patterns. Such an increase in efficiency may be one of the principle reasons that feedback motifs are enriched within developmental networks.

**Acknowledgements.** Research reported in this publication was supported by the National Institute of General Medical Sciences of the National Institutes of Health under Award Number P50GM076547. Thanks to Ilya Shmulevich for helpful discussions. The content is solely the responsibility of the authors and does not necessarily represent the official views of the National Institutes of Health.

# References

1. Lander, A.D.: Morpheus unbound: reimagining the morphogen gradient. Cell **128**(2), 245–256 (2007). http://dx.doi.org/10.1016/j.cell.2007.01.004
2. Lander, A.D.: Pattern, Growth, and Control. Cell **144**(6), 955–969 (2011). http://dx.doi.org/10.1016/j.cell.2011.03.009
3. Flann, N.S., Mohamadlou, H., Podgorski, G.J.: Kolmogorov complexity of epithelial pattern formation: the role of regulatory network configuration. Biosystems **112**(2), 131–138 (2013). http://dx.doi.org/10.1016/j.biosystems.2013.03.005

4. Davidson, E.H.: Emerging properties of animal gene regulatory networks. Nature **468**(7326), 911–920 (2010). http://dx.doi.org/10.1038/nature09645
5. Shen-Orr, S.S., Milo, R., Mangan, S., Alon, U.: Network motifs in the transcriptional regulation network of Escherichia coli. Nat. Genet. **31**(1), 64–68 (2002). http://dx.doi.org/10.1038/ng881
6. Milo, R., Shen-Orr, S., Itzkovitz, S., Kashtan, N., Chklovskii, D., Alon, U.: Network motifs: simple building blocks of complex networks. Science **298**(5594), 824–827 (2002). http://dx.doi.org/10.1126/science.298.5594.824
7. Barabasi, A.-L., Oltvai, Z.N.: Network biology: understanding the cell's functional organization. Nat. Rev. Genet. **5**(2), 101–113 (2004). http://dx.doi.org/10.1038/nrg1272
8. Ghaffarizadeh, A., Flann, N., Podgorski, G.: Multistable switches and their role in cellular differentiation networks. BMC Bioinf. **15**(Suppl. 7), S7+ (2014). http://dx.doi.org/10.1186/1471-2105-15-s7-s7
9. Mazumdar, A., Mazumdar, M.: How one becomes many: blastoderm cellularization in Drosophila melanogaster. BioEssays : News Rev. Mol. Cell. Dev. Biol. **24**(11), 1012–1022 (2002). http://dx.doi.org/10.1002/bies.10184
10. Goodyear, R., Richardson, G.: Pattern formation in the basilar papilla: evidence for cell rearrangement. J. Neurosci. Official J. Soc. Neurosci. **17**(16), 6289–6301 (1997). http://view.ncbi.nlm.nih.gov/pubmed/9236239
11. Podgorski, G.J., Bansal, M., Flann, N.S.: Regular mosaic pattern development: a study of the interplay between lateral inhibition, apoptosis and differential adhesion. Theor. Biol. Med. Model. **4**, 43+ (2007). http://dx.doi.org/10.1186/1742-4682-4-43
12. Kolmogorov, A.N.: Three approaches to the quantitative definition of information. Prob. Inf. Transm. **1**, 1–7 (1965)
13. Kauffman, S.A.: Metabolic stability and epigenesis in randomly constructed genetic nets. J. Theor. Biol. **22**(3), 437–467 (1969). http://view.ncbi.nlm.nih.gov/pubmed/5803332
14. Eglen, S.J., Willshaw, D.J.: Influence of cell fate mechanisms upon retinal mosaic formation: a modelling study. Development **129**(23), 5399–5408 (2002). http://view.ncbi.nlm.nih.gov/pubmed/12403711
15. Kim, M.S., Kim, D., Kim, A., Lander, A.D., Cho, K.H.: Spatiotemporal network motif reveals the biological traits of developmental gene regulatory networks in Drosophila melanogaster. BMC Syst. Biol. **6**(1), 31+ (2012). http://dx.doi.org/10.1186/1752-0509-6-31
16. Clune, J., Mouret, J.-B., Lipson, H.: The evolutionary origins of modularity. Proc. R. Soc. B: Biol. Sci. **280**(1755), 20122863 (2013). http://dx.doi.org/10.1098/rspb.2012.2863
17. Lorenz, D.M., Jeng, A., Deem, M.W.: The emergence of modularity in biological systems. Phys. Life Rev. **8**(2), 161–162 (2012). http://arxiv.org/abs/1204.5999
18. Kim, J.-R., Yoon, Y., Cho, K.-H.H.: Coupled feedback loops form dynamic motifs of cellular networks. Biophys. J. **94**(2), 359–365 (2008). http://dx.doi.org/10.1529/biophysj.107.105106
19. Kalir, S., Mangan, S., Alon, U.: A coherent feed-forward loop with a SUM input function prolongs flagella expression in Escherichia coli. Mol. Syst. Biol. **1**(1), msb4 100 010-E11–msb4 100 010-E16 (2005). http://dx.doi.org/10.1038/msb4100010
20. Alon, U.: Network motifs: theory and experimental approaches. Nat. Rev. Genet. **8**(6), 450–461 (2007). http://dx.doi.org/10.1038/nrg2102

21. Singh, H., Khan, A.A., Dinner, A.R.: Gene regulatory networks in the immune system. Trends Immunol. **35**(5), 211–218 (2015). http://dx.doi.org/10.1016/j.it.2014.03.006
22. Levine, M., Davidson, E.H.: Gene regulatory networks for development. Proc. Nat. Acad. Sci. U.S.A. **102**(14), 4936–4942 (2005). http://dx.doi.org/10.1073/pnas.0408031102
23. Swiers, G., Patient, R., Loose, M.: Genetic regulatory networks programming hematopoietic stem cells and erythroid lineage specification. Dev. Biol. **294**(2), 525–540 (2006). http://dx.doi.org/10.1016/j.ydbio.2006.02.051
24. Ma, H.-W., Kumar, B., Ditges, U., Gunzer, F., Buer, J., Zeng, A.-P.: An extended transcriptional regulatory network of Escherichia coli and analysis of its hierarchical structure and network motifs. Nucleic Acids Res. **32**(22), 6643–6649 (2004). http://dx.doi.org/10.1093/nar/gkh1009
25. Kauffman, S.: The Origins of Order: Self-Organization and Selection in Evolution. Oxford University Press, New York (1993). http://www.ncbi.nlm.nih.gov/pmc/articles/PMC1226010/
26. Galas, D.J., Nykter, M., Carter, G.W., Price, N.D., Shmulevich, I.: Biology information as set-based complexity. IEEE Trans. Inf. Theory **56**(2), 667–677 (2010). http://dx.doi.org/10.1109/TIT.2009.2037046
27. Ignac, T., Sakhanenko, N., Galas, D.: Relations between the set-complexity and the structure of graphs and their sub-graphs. EURASIP J. Bioinf. Syst. Biol. **2012**(1), 1–10 (2012). http://dx.doi.org/10.1186/1687-4153-2012-13
28. Carter, G.W., Rush, C.G., Uygun, F., Sakhanenko, N.A., Galas, D.J., Galitski, T.: A systems-biology approach to modular genetic complexity. Chaos: Interdisc. J. Nonlinear Sci. **20**(2), 026 102+ (2010). http://dx.doi.org/10.1063/1.3455183
29. Chen, X., Francia, B., Li, M., McKinnon, B., Seker, A.: Shared information and program plagiarism detection. IEEE Trans. Inf. Theory **50**(7), 1545–1551 (2004). http://dx.doi.org/10.1109/tit.2004.830793

# Harmonic Versus Chaos Controlled Oscillators in Hexapedal Locomotion

Luis A. Fuente[1]([envelope]), Michael A. Lones[2], Nigel T. Crook[1], and Tjeerd V. Olde Scheper[1]

[1] Department of Computing and Communication Technologies, Oxford Brookes University, Oxford OX3 0BP, UK
{lfuente-fernandez,ncrook,tvolde-scheper}@brookes.ac.uk
[2] Department of Computer Science, Heriot-Watt University, Edinburgh EH14 4AS, Scotland, UK
M.Lones@hw.ac.uk

**Abstract.** The behavioural diversity of chaotic oscillator can be controlled into periodic dynamics and used to model locomotion using central pattern generators. This paper shows how controlled chaotic oscillators may improve the adaptation of the robot locomotion behaviour to terrain uncertainties when compared to nonlinear harmonic oscillators. This is quantitatively assesses by the stability, changes of direction and steadiness of the robotic movements. Our results show that the controlled Wu oscillator promotes the emergence of adaptive locomotion when deterministic sensory feedback is used. They also suggest that the chaotic nature of chaos controlled oscillators increases the expressiveness of pattern generators to explore new locomotion gaits.

## 1 Introduction

Living organisms use distinctive locomotive abilities to interact adaptively with their environment. Central Pattern Generators (CPGs) are a popular approach inspired by the networks of neurons in spinal cords of vertebrates and invertebrate thoracic ganglia to develop accurate neuro-mechanical representations of locomotive behaviours in legged robots. Whilst traditional CPGs are often dynamically represented by coupled of harmonic oscillators [1,2], recent studies envisage CPGs as a collection of coupled chaotic oscillators [4,5] which mutually self-regulate according to environmental fluctuations and local influences.

Although the consideration of sensory information is not essential in the generation of synchronous pattern of motion, it increases the adaptability and robustness of control system in unsteady real environments. Orchestrating environmental signals to achieve efficient locomotion is a difficult task, specially when these signals come from potentially noise sources (i.e. robotic sensors). However, this is a task at which signalling networks are evidently good at solving in a biological contexts. We collectively refer to computational analogies of cellular signalling networks as Artificial Signalling Networks (ASNs) [8,9].

© Springer International Publishing Switzerland 2015
M. Lones et al. (Eds.): IPCAT 2015, LNCS 9303, pp. 114–127, 2015.
DOI: 10.1007/978-3-319-23108-2_10

In this article, we explore whether adaptability in multi-legged robots can be reinforced by the structural and dynamical properties of chaotic oscillators when compared to harmonic oscillators. Particularly, we aim to investigate whether the use of chaos controlled oscillators offer more flexibility in the generation of effective patterns of coordination with adequate sensory feedback loops. To do so, a two-layered architecture is used to map environmental signals into adaptive locomotive trajectories. The upper layer consists of a collection of interconnected ASNs. It receives real-time positional information and produces a set of environmental-based control directives, which collectively alter in a deterministic manner the dynamics of a CPG composed of either Hopf or controlled Wu oscillators in the bottom layer. Both oscillators exhibit a stable limit circle, but otherwise they lie at opposite ends of the dynamical spectrum: the former is a single steady state system and the latter shows ordered and chaotic behaviours depending on its governing parameters. We use an evolutionary algorithm to optimise computational analogies of signalling networks that, when stimulated with sensory feedback, tune the CPG's oscillatory trajectories appropriately. A simulated version of the T-Hex robot is used to evaluate the performance of both oscillators in a challenging environment.

The rest of the article is organised as follows: Sect. 2 introduces the oscillators addressed in this article, Sect. 3 describes the CPGs, Sect. 5 describes our methodology, Sect. 4 presents signalling networks, Sect. 6 presents results and [8] Sect. 7 concludes.

## 2   Central Pattern Generators

Central Pattern Generators (CPGs) are a common way of modelling and generating locomotive gaits. Whilst they are often implemented using biologically-motivated models such as feedforward neural networks [13] and artificial biochemical networks [12], they can also be considered as systems of coupled nonlinear oscillators. This kind of CPG favours distributed control approaches, leg synchronisation and the modulation of locomotion by simple control signals. Its simple structure also eases the integration of sensory information when CPGs made of coupled nonlinear oscillators are applied to the control of robotic locomotion.

### 2.1   Hopf Oscillator

The Hopf oscillator (see Fig. 1) is a single steady state dynamical system that exhibits harmonic oscillation [14]. It is defined by the following two differential equations:

$$\dot{y} = \alpha(\mu - r^2)y - \omega z \qquad\qquad \dot{z} = \beta(\mu - r^2)z - \omega y \qquad (1)$$

where $(y, z) \in \mathbb{R}^2$ are the state variables, $r = \sqrt{y^2 + z^2}$, $A = \sqrt{\mu}$ is the oscillation amplitude, $\omega$ is the oscillation frequency and $\alpha$ and $\beta$ are positive constants that determine its convergence rate to the limit circle. From our perspective, the Hopf oscillator has three prominent benefits. First, it is able to generate smooth, stable

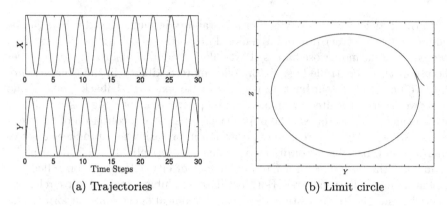

(a) Trajectories                    (b) Limit circle

**Fig. 1.** Outputs of the Hopf oscillator with $\mu = 1$, $\omega = 2$, $\alpha = 5$ and $\beta = 50$ and a phase portrait of its stable limit circle attractor.

and cyclic trajectories in the presence of small perturbations. Second, its output can be exclusively modulated by changing its frequency $\omega$ and amplitude $\mu$, whilst preserving the other parameters. This separation eases its optimisation by evolutionary algorithms [15]. Third, it eases the development of coupling terms in an analytical way.

## 2.2    Wu Oscillator

The Wu oscillator (see Fig. 2(a, b)) is a four-dimensional autonomous dynamical system able to exhibit a large variety of dynamical states [16]. It is defined by the following set of differential equations:

$$\dot{x} = a(y - x) + eyz - kw \qquad \dot{y} = cx - dy - xz \qquad (2)$$
$$\dot{z} = xy - bz \qquad \dot{w} = ry + fyz \qquad (3)$$

where $(x, y, z, w) \in \mathbb{R}^4$ and $a$, $b$, $c$, $d$, $e$, $f$, $k$ and $r$ are all real constants. This oscillator is adopted for several reasons. Its four nonlinear terms help to rapidly propagate small alterations across its variables and, when coupled, throughout neighbouring oscillators. The Wu oscillator is also able to self-regulate and self-sustain its internal dynamics by adjusting its amplitude and frequency in response to external signals. When coupled, the system's overall state can only be deduced from the interactions amongst individual oscillators.

**Trajectory Stabilisation.** Individual chaotic trajectories of the Wu oscillator are stabilised using the Rate Control of Chaos (RCC) method [17]. Unlike other chaos control strategies, the RRC method does not require any *a priori* knowledge about the presence of unstable periodic trajectories in a chaotic system. This approach relies on the expansion rate of an oscillator away from its trajectory to apply a small scale into its governing variables proportion to the

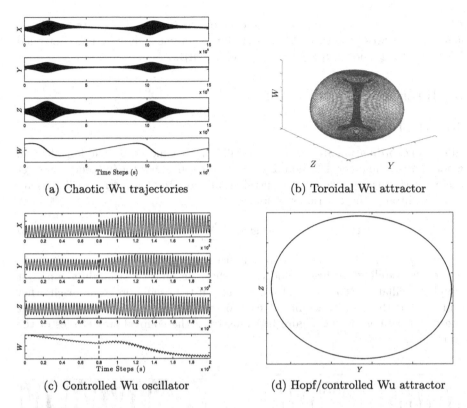

(a) Chaotic Wu trajectories

(b) Toroidal Wu attractor

(c) Controlled Wu oscillator

(d) Hopf/controlled Wu attractor

**Fig. 2.** Outputs of the Wu and controlled Wu oscillators. In both cases, the state parameters are $a = 56$, $b = 16$, $c = 49$, $d = 9$, $k = 8$, $e = 30$, $f = 40$ and $r = -1943$. Using this configuration the Wu oscillator exhibits an aperiodic chaotic behaviour (a) with all its trajectories converging to a toroid attractor (b). When the RCC method is active at $t = 0.8 \times 10^4$ (vertical dotted line), the oscillator's unstable and aperiodic trajectories (c) become harmonic with constant frequency and amplitude after an initial transient time (wandering period). (d) illustrates the stable limit circle attractor of the controlled Wu oscillator.

divergence rate. Thus, it is possible to reduce the chaotic nature of the oscillator but preserving its chaotic properties. The extend of the perturbation can be calculated by determining the current proportion the variable occupies in its space. Within a robotic locomotion context, this also allows the modulation of the state of the robot based uniquely on the CPG's local influences and sensory feedback without any explicit knowledge of the robot's surrounding environment. The equations of the chaos controlled Wu oscillator are as follows:

$$\dot{x} = a(y - x) + e\sigma_x yz - kw \qquad \dot{y} = cx - dy - \sigma_y xz \qquad (4)$$
$$\dot{z} = \sigma_z xy - bz \qquad \dot{w} = ry + f\sigma_w yz \qquad (5)$$

where $\sigma_x$, $\sigma_y$, $\sigma_z$ and $\sigma_w$ are the rate control functions and regulate the divergence rate between the variables in each of the nonlinear terms[1]. The controlled Wu oscillator is shown in Figures (see Fig. 2(c, d)).

## 3   Interlimb Coordination

### 3.1   Hopf Oscillator

Interlimb coordination amongst Hopf oscillators is achieved by coupling them in a non-diffusive manner. It is non-diffusive in the sense that the influence amongst oscillator is constant over time. Contralateral and lateral adjacent oscillators are coupled through the $\dot{z}$ variable as follows:

$$\dot{z} = \beta(\mu - r_i^2)z_i - \omega_i x_i + \sum k_{ij}(z_i + \lambda_{ij}z_j) \tag{6}$$

where $i, j = 1\ldots6$ are the oscillator indices, $k_{ij}$ is the diffusive coupling and $\lambda_{ij}$ is the coupling coefficient and establishes the phase relationships amongst coupled oscillators. The value of $\lambda$ is set to 1 if the oscillators excite each other and to -1 if the oscillators inhibit each other. We chose a coupling such that the tripod gait is stable. Figure 3(a) illustrates the coupled trajectories of the Hopf CPG.

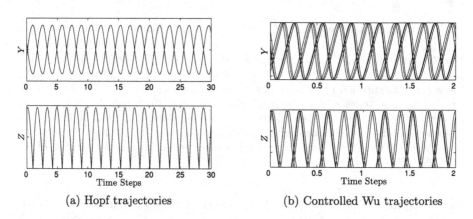

|                          |                               |
| :----------------------: | :---------------------------: |
| (a) Hopf trajectories    | (b) Controlled Wu trajectories |

**Fig. 3.** Coupled $y$ (upper plot) and $z$ (bottom plot) trajectories of two CPGs composed of six Hopf and controlled Wu oscillators. For the Hopf CPG, trajectories are obtained for $k_{ij} = 1$. For the Wu CPG, trajectories are obtained using $\sigma = 0.0118$, $\tau_{12} = 1.0$, $\tau_{13} = 1.8$, $\tau_{24}$, $\tau_{34} = \tau_{35} = 1.1$, and $\tau_{56} = 1.8$. We exploit the fact that $\tau_{ij} = \tau_{ji}$. Although Wu CPG trajectories are not completely synchronised, their minimal phase difference does not affect the performance of the control system and allows preserving stable forward locomotion.

---

[1] Refer to [17] for additional insight about controlling unstable trajectories using the RRC method.

## 3.2    Chaos Controlled Wu Oscillator

Chaos controlled Wu oscillators are coupled in a non-diffusive manner with time delay feedback. Contralateral and lateral adjacent oscillators are coupled through the $\dot{x}$ variable as follows:

$$\dot{x}_i = a(y_i - x_i) + e\sigma(y_i, z_i)y_i z i - k w_i + \sum \tau_{ij}(x_j(t - \sigma) - x_i) \qquad (7)$$

where $i, j = 1 \ldots 6$ are the oscillator indices, $\sigma$ is the time delay and $\tau_{ij}$ is the coupling coefficient and defines the effect of the $i$th on the $j$th oscillator thereby establishing their phase relationship. The $y$ and $z$ trajectories of the controlled Wu oscillators are chosen to describe the motion of each leg because they increase stability when the robot moves on a flat surface. The coupling amongst trajectories also matches the tripod gait. Figure 3(b) illustrates the coupled trajectories of the Wu CPG.

# 4    Coupled Artificial Signalling Network

An Artificial signalling Network (ASN) is an enzyme-mediated abstraction of a cellular signalling process. Cellular signalling is the biological mechanism by which cells interact with other cells and their environment [7]. Broadly speaking, cellular signalling is a chain of events that, triggered by an extracellular signal, induces an adaptive cellular response. It begins with the binding of certain messengers and their later diffusion inside the cell. Such messengers then spread across the cell using signalling pathways until they reach the nucleus where they regulate gene expression and lead to the process by which a change in the cellular activity can be achieved.

Formally, an ASN is an indexed set of enzyme-analogous elements $E$ and a set of continuous-valued biochemical reactions. Each $e_i \in E$ has a set of substrates $s_i$, a product concentration $p_i$ and a regulatory function $f_i$. Substrate concentrations are mapped to product concentrations using the probabilistic Michaelis-Menten function which was previously shown to lead to the best performances when evolving ASNs capable of controlling trajectories in a prescribed manner [10]. The execution of the ASN starts with the random initialisation of the concentrations ($s_i$ and $p_i$). External inputs are delivered to the network by the substrate concentration of nominated enzymes. At each time step, each enzyme $e_i$ applies its regulatory function $f_i$ to the current concentration of its substrates $s_i$ to determine the new concentration of its product $p_i$. After iterating the network a specific number of times $t_S$, the outputs are extracted from the final product concentration of designated enzymes.

Biological responses to sensory information are often the result of the interaction of multiple pathways. Motivated by this observation, a number of authors have investigated the coupling of computational architectures based on models of interacting biochemical networks [11,12]. This article focuses on the interaction amongst signalling pathways since this is the principal process through which biological organisms handle environmental information. One of the main

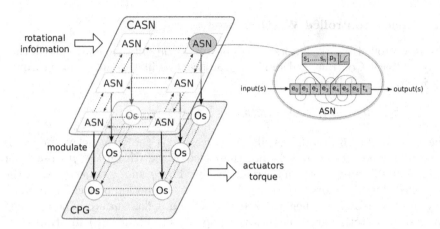

**Fig. 4.** Overview of the locomotive control system and coupling topology. Sensory feedback is delivered to an upper layer composed of six interconnected signalling networks, one per leg. Coupled networks comprise 10 enzyme-analogous elements. Gait trajectories are generated in the bottom layer, which contains a pattern generator made of six nonlinear oscillators. This layer, whose connectivity mirrors the upper layer, receives control directives that modulate the oscillatory trajectories, also one per leg. Finally, these are transformed into actuator positions.

mechanisms that allows the exchange of information between interconnected signalling networks is crosstalk. In [8], we introduced a simple model of crosstalk within a connectionist architecture called a Coupled Artificial Signalling Network (CASN). Mimicking the structure of a signalling network, different types of external inputs are delivered to different sub-networks. These sub-networks (which are comparable to ASNs) do not have explicit interconnections, but they do contain crosstalk nodes which permit the exchange of information using a simple regulatory function. In this article, CASNs are optimised using evolutionary algorithms in order to transform sensory data into deterministic control directives that when applied to the governing parameters of each oscillator in the CPG alter its dynamics and elicit adaptive modifications in the locomotion pattern. From a locomotive perspective, coupling adjacent ASNs also favours the synchronisation of CASN-modulated oscillatory trajectories [8] (Fig. 4).

## 5    Controlling Legged Robot Locomotion

A simulated model (see Fig. 5) of the commercial T-Hex robot is used to evaluate the expressiveness of the Hopf and the Wu CPGs. The T-Hex is a 24-DoF hexapedal robot manufactured by Lynxmotion [18]. It has four joints per leg connected by actuators at the corners. The robot initially walks using the tripod gait, which is described by the moving of three legs simultaneously in each step. Its limited adaptability on irregular surfaces is exploited to determine the capacity of the CPGs to deliver reactive locomotion using local sensory feedback. The

**Fig. 5.** Simulated T-Hex robot in open dynamics engine.

robot is simulated using the Open Dynamics (ODE) physics engine with a step side of $\Delta_t = 0.01$ s, friction of 200 N, CFM (an ODE parameter) of $10^{-5}$ and standard gravity. Actuators have a maximum angular velocity of 4 ms$^{-1}$ and a maximum torque of 70 Nm. Their movements are limited in both the z-axis plane for the coxa joint and the x-axis for the femur, tibia and tarsal joints, to a maximum rotation of 90° and a minimum rotation of $-90°$. These values are sufficient to simulate the characteristics of the physical T-Hex [3,10]. The CASN is executed every 20 simulation steps.

### 5.1 Gait Generation

The task is to evolve a CASN capable of generating control signals that when applied to the oscillators' governing parameters originate different outputs in the CPG, i.e. patterns of coordination which would cause the robot to displace away from its starting point. The aim is to measure qualitatively whether chaotic properties of the controlled Wu oscillators are sufficient to generate adaptive patterns of movement in response to deterministic sensory feedback when compared with the harmonic Hopf oscillator. The movement of each leg is controlled. The CASN consists of six individual ASN uniquely coupled to its adjacent ASN using a fixed crosstalk rate of 0.5. Each ASN is immediately connected with its corresponding bottom-level oscillator, whose output dictates the gait trajectory its matching leg. The controller fitness is the Euclidean distance minus the lateral displacement walked by the robot within an evaluation period of 4000 simulation steps. Both the Hopf and the Wu CPGs are randomly initialised and numerically integrated using the fifth-order Dormand-Prince method with step sizes of $\Delta_{t_H} = 0.01$ and $\Delta_{t_W} = 0.00001$ respectively. The selected $y$ and $z$ trajectories are scaled to a maximum height of 40 mm and a maximum length of 30 mm respectively, and sampled with a rate of $s_r = \pi/4$ ($\approx 40t_H$ and $\approx 380t_W$). The population size is 200, with a generation limit of 100.

The rotational readings along the three Cartesian axes of each leg with respect to the centre of the robot represent the inputs of each ASN. They are the easiest feedback that gives actual insight into the robot stability and terrain features. Our objective is to calculate the rate of control needed to stabilise

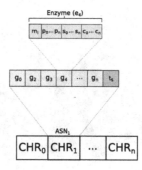

**Fig. 6.** Linear encoding of the CASN network used by the evolutionary algorithm, also showing how individual ASN and enzymes are represented.

the robot whilst promoting the emergence of different patterns of movement. Rotational values are matched to the $[-\pi/3, \pi/3]$ interval, linearly scaled to the concentration range of $[0, 1]$, and delivered to the ASN through its substrate concentrations. Values out of the rotation interval indicate that the robot has fallen. For the Hopf CPG, each ASN has two outputs which match the $\omega$ and $\mu$ Hopf parameters and modulate them in the intervals $[0, 4]$ and $[0, 8]$ respectively. For the Wu CPG, each ASN has also two outputs which match the $c$ and $b$ Wu constant and modulate them in the intervals $[44, 54]$ and $[11, 21]$ respectively. In both cases, the outputs are in the range of $[0, 1]$. The behaviour of the Hopf and controlled Wu pattern generators is evaluated on an uneven terrain consisting of a starting zone for the robot and a randomly generated uneven terrain, which comprises a mesh of 500 boxes with randomly chosen heights. Each box has a side of 20 mm and incremental height between 20 mm and 45 mm. Values over these thresholds prevent the robot from moving forwards.

### 5.2  Evolving Coupled Artificial Signalling Networks

CASNs are evolved using a standard generational evolutionary algorithm with tournament selection (size = 4), uniform crossover (rate = 0.3), and point mutation (rate = 0.05). Each individual in the population is encoded as an indexed sequence of chromosomes, each of which represents an ASN (see Fig. 6). A multi-chromosomal representation favours the evolution of problems with growing complexity and increases modularity [19]. Signalling networks are encoded as a set of 10 indexed genes followed by timing information. Crossover points lie between gene boundaries and chromosome shuffling is not permitted. Inputs and outputs ($s_i$ and $p_i$) are represented by their absolute indices. Mutation is restricted to the set of operations in [20] to embrace biochemical plausibility in the evolution of enzymatic graphs. Chemical concentrations and function parameters are represented using floating-point values and mutated using a Gaussian function centred around the current value.

## 6   Results

The results show that CASNs can be used to deliver reactive control directives to CPGs with distinctive dynamics. While the robots actuated by the Hopf CPG exhibits an average walked distance of 67 cm (s.d. 21 cm), the robots actuated by the controlled Wu CPG exhibit an averaged walked distance of 58 cm (s.d. 12 cm). However, the performance of the Hopf CPG decreases drastically when the robot steps over the uneven terrain. Figure 7 shows the trajectories of each robot controlled during the evaluation time. Notably, the trajectories of the Hopf GPG start diverging apart as the robot steps over the test area. The results is that, in some runs, the robot laterally exits the uneven section of the terrain, giving rise to the furthest walking distances. Particularly, Hopf CPG-based robots experience noticeable difficulties to move when the terrain's height is 20 mm and nearly do not move when the terrain's height is up to 25 mm. On the contrary, the trajectories of the Wu CPG remain grouped until the terrain's height is approximately 25 mm, the movement at which they progressively diverge always within the test arena. This may suggests that the controlled Wu CPGs exhibit a better regulatory capacity as the intensity of the sensory feedback increases.

Figure 8 illustrates the overall rotation of the robot body along the $x$-axis (a), (b) and $y$-axis (c), (d) for both CPGs. Attending to the number CASN executions, it is also noticeable that robots actuated by the Hopf CPG reach the uneven part of the terrain quicker than the ones actuated by the controlled Wu CPG. However, the controlled Wu CPGs lead to more stable and steady locomotive movements on the flat surface despite that the Wu oscillators are not perfectly synchronised. A possible explanation is that the coupling error may promote stability since representative sensory feedback is received to the CASN

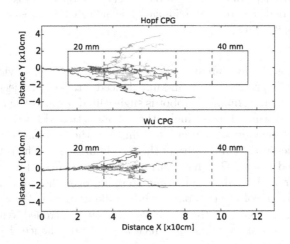

**Fig. 7.** Trajectories of each of the 10 evolved robot controllers when moving forwards in the test arena. The box represented with a solid line illustrates the bounds of the uneven terrain. Its initial maximum height is 20 mm and increases in steps of 5 mm after each horizontal dashed line up to 40 mm.

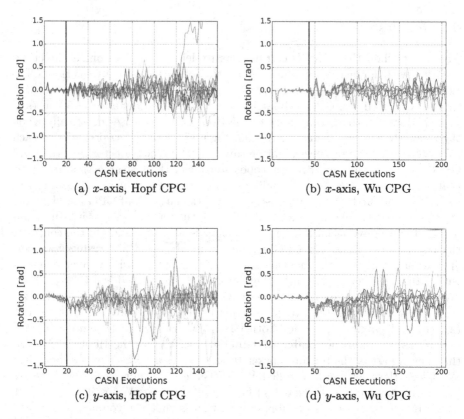

**Fig. 8.** Rotations of the robot body along the $x$- and $y$-axes for the Hopf CPGs (left plot) and the controlled Wu CPGs (right plot) every time the CASN is executed during the evaluation period. The vertical solid black line indicate the movement at which the robot reaches the uneven terrain.

from the starting of the evaluation period. Further, it is also evident that the Hopf CPGs readily produce sharp and unbalanced rotational trajectories. As can be seen in Fig. 8(c) the simulated robot is slight tilted towards the right throughout the evaluation time. Interestingly, the robots actuated by Hopf CPGs have less difficulty stepping over the rough terrains, but they exhibit more fuzzy rotational trajectories as the complexity of the terrain increases. The robots actuated by the controlled Wu CPG behave in an opposite manner. They show abrupt changes in their walking direction while stepping over the uneven terrain, but they manage to control their stability while walking over it. We can hypothesis that this is a consequence of the flexibility of chaos controlled Wu oscillator to explore new pattern of coordination which appears to enhance the stabilisation of the robot in different terrains.

Adaptive locomotion is inherently difficult to analyse. This is because locomotive patterns of coordination not only depend on sensory feedback but also on the local interactions amongst the oscillators in a CPG. However, techniques to

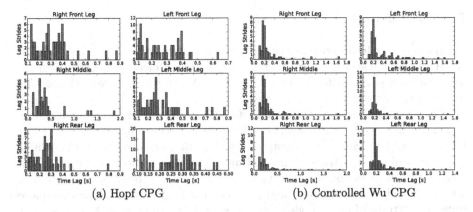

(a) Hopf CPG          (b) Controlled Wu CPG

**Fig. 9.** Example of the distributions of the time interval between consecutive touch-downs for each leg of the simulated robot when actuated by a Hopf CPG (left plot) and a controlled Wu CPG (right plot) during the evaluation time.

study the temporal distribution of the stepping patterns can be used to provide some insight into the behaviour of the CPG [21]. Figure 9 exemplifies the patterns of movement of two runs as the distribution of the time interval between consecutive touchdowns for each leg of the hexapod robot. In general, we found that the Wu CPGs have a clearly defined stepping pattern, which suggests that some self-induced patterns of motion may exist for this type of controller.

Figure 9(a) depicts histograms whose shape resembles a Gaussian-like distribution with a time lag mean steadily centred at 0.2 s. This is indicative that the robot changes its initial fast tripod gait to another locomotive gait with smaller stance phase and longer swinging phase. This is perhaps surprising since there is not explicit coordination between oscillators during the evaluation time apart form the initial tripod coupling. In addition, this also suggests that anti-phase synchronisation between legs arises despite the dissimilarity of the sensory feedback. To a certain extent, this is also a consequence of the RCC method which regulates the effect that external perturbations have in the controlled Wu oscillator by allowing it to migrate to another area in the state space whilst preserving the integrity of its limit cycle. On the contrary, the Hopf CPGs produce rather irregular patterns of coordination (see Fig. 9(b)) in which no real synchronisation occurs amongst oscillators. As a consequence, the robot either preserves its initial tripod coupling while exploring gaits around it, or explores new patterns of coordination but it is unable to preserve any of them. Nonetheless, the Hopf CPG is sufficient to induce reactive locomotive patterns when the irregularities of the terrain are not severe. This also explains the differences in performance seen in Figs. 7 and 8, with the controlled Wu CPGs leading to more stable solutions whose patterns of coordination are more likely to adapt to changing environments.

# 7  Conclusions

In this paper, we have compared the performance of two different CPGs based on the Hopf and Wu systems in a rough terrain with incremental difficulty. Sensory information was added to the systems using CASNs, which were optimised for each controller using evolutionary algorithms.

The robots actuated by the Hopf CPG achieve the furthest walking distances and show the fastest locomotive paces, but also exhibit a remarkable lack of stability on rough terrains. The robots actuated by the controlled Wu CPG exhibit more deterministic patterns of locomotion. Likewise, the intrinsic nature of the controlled Wu oscillator allows the emergence of new patterns of adaptation in a reactive and efficient manner and increases the flexibility of CPGs to explore different patterns of motion when compared with the Hopf oscillator. Overall, it appears that the chaotic nature of the chaotic rate controlled Wu CPG enhances the development of adaptive and robust behaviours using sensory feedback.

In future work, we plan to investigate the importance of sensory feedback in the generation of differential patterns of motion and to evaluate the performance of both CPGs in the physical T-Hex and alternative surfaces.

# References

1. Pina Fhilo de, A.C., Dutra, S.M., Raptopoulos, L.S.C.: Modelling of a bipedal robot using mutually coupled oscillators. Biol. Cybern. **92**(1), 1–7 (2005)
2. Righetti, L., Ijspeert, A.J.: Design methodologies for central pattern generators: an application to crawling humanoids. In: Proceedings of Robotics: Science and Systems, pp. 191–198 (2006)
3. Fuente, L.A., Lones, M.A., Turner, A.P., Caves, L.S., Stepney S., Tyrrell A.M.: Adaptive robotic gait control using coupled artificial signalling networks, hopf oscillators and inverse kinematics. In: 15th IEEE Congress on Evolutionary Computation, pp. 1435–1442 (2013)
4. Kuniyoshi, Y., Suzuki, S.: Dynamic emergence and adaptation of behaviour through embodiment as coupled chaotic fields. In: IEEE/RSJ International Conference on Intelligent Robots and Systems, vol. 2, pp. 2042–2049 (2004)
5. Matthey, L., Righetti, L., Ijspeert, A.J.: Experimental study of limit cycle and chaotic controllers for the locomotion of centipede robots. In: IEEE/RSJ International Conference on Intelligent Robots and Systems, pp. 1860–1865 (2008)
6. Fasca, M., Arena, P., Fortuna, L.: Bio-inspired Emergent Control of Locomotion Systems. World Scientific Series of Nonlinear Science A, vol. 48. World Scientific, London (2004)
7. Cell signalling: $H_2O_2$ a necessary evil for cell signalling. Science **312**, 1882–1883 (2006)
8. Fuente, L.A., Lones, M.A., Turner, A.P., Stepney, S., Caves, L.S., Tyrrell, A.M.: Evolved artificial signalling networks for the control of a conservative complex dynamical system. In: Lones, M.A., Naef, F., Smith, S.L., Teichmann, S., Trefzer, M.A., Walker, J.A. (eds.) IPCAT 2012. LNCS, vol. 7223, pp. 38–49. Springer, Heidelberg (2012)

9. Fuente, L.A., Lones, M.A., Turner, A.P., Caves, L.S., Stepney, S., Tyrrell, A.M.: Computational models of signalling networks for non-linear control. BioSystems **122**(2), 122–130 (2013)
10. Fuente, L.A.: A decentralised control architecture: coupled artificial signalling networks. Ph.D. thesis (2014)
11. Lones, M.A., Caves, L.S., Stepney S., Tyrrell A.M.: Controlling legged robots with coupled artificial biochemical networks. In: ECAL 2011, pp. 465–472. MIT Press (2011)
12. Lones, M.A., Fuente, L.A., Turner, A.P., Caves, L.S., Stepney, S., Tyrrell, A.M.: Artificial biochemical networks: evolving dynamical systems to control dynamical systems. IEEE Trans. Evol. Comput. **18**(2), 145–166 (2014)
13. Moioli R.C., Vargas, P.A., Husbands, P.: A multiple hormone approach to homeostatic control of conflicting behaviours in an autonomous mobile robot. In: IEEE 11th Congress on Evolutionary Computation, pp. 47–54 (2009)
14. Righetti, L., Ijspeert, A.J.: Pattern generators with sensory feedback for the control of quadruped locomotion. In: IEEE International Conference on Robotic and Automation, pp. 819–824 (2008)
15. Wright, J., Jordanov, I.: Intelligent approaches in locomotion. In: The 2012 International Joint Conference in Neural Networks (IJCNN 2012), pp. 1–8 (2012)
16. Wu, W., Zengqiang, C., Zhuzhi, Y.: The evolution of a novel four-dimensional autonomous system: among 3-torus, limit cycle, 2-torus, chaos and hyper-chaos. Chaos Solitions Fractals **39**(5), 2340–2356 (2009)
17. Scheper, T.O.: Why metabolic systems are rarely chaotic? BioSystems **94**(1–2), 145–152 (2008)
18. Lynxmotion Ltd.: 4-dof t-hex combo kit for bot board/ssc-32/bap28. http://www.lynxmotion.com/c-151-t-hex-4-dof.aspx
19. Mayer, H.A., Spitzlinger, M.: Multi-chromosomal representations and chromosome shuffling in evolutionary algorithms. In: The 2003 Congress on Evolutionary Computation, vol. 2, pp. 1145–1149 (2003)
20. Ziegler, J., Banzhaf, W.: Evolving control metabolisms for a robot. Artif. Life **7**, 171–190 (2001)
21. Abourachid, A., Herbin, A., Hacker, R., Maes, L.: Experimental study of coordination patterns during unsteady locomotion in mammals. J. Exp. Biol. **210**, 366–372 (2007)

# Biochemical Regulatory Networks

# Switching Gene Regulatory Networks

Yoli Shavit[1,2], Boyan Yordanov[2], Sara-Jane Dunn[2],
Christoph M. Wintersteiger[2], Youssef Hamadi[2], and Hillel Kugler[2,3(✉)]

[1] University of Cambridge, Cambridge, UK
[2] Microsoft Research, Cambridge, UK
hkugler@outlook.com
[3] Bar-Ilan University, Ramat Gan, Israel

**Abstract.** A fundamental question in biology is how cells change into specific cell types with unique roles throughout development. This process can be viewed as a program prescribing the system dynamics, governed by a network of genetic interactions. Recent experimental evidence suggests that these networks are not fixed but rather change their topology as cells develop. Currently, there are limited tools for the construction and analysis of such self-modifying biological programs. We introduce *Switching Gene Regulatory Networks* to enable the modeling and analysis of network reconfiguration, and define the synthesis problem of constructing switching networks from observations of cell behavior. We solve the synthesis problem using Satisfiability Modulo Theories (SMT) based methods, and evaluate the feasibility of our method by considering a set of synthetic benchmarks exhibiting typical biological behavior of cell development.

**Keywords:** Gene regulatory networks (GRNs) · Boolean networks · Biological Modeling · Satisfiability Modulo Theories (SMT) · Synthesis · Self-modifying code

## 1 Introduction

The cell is a fundamental unit of biological systems. Many aspects of cellular function are interpreted as the consequence of a series of genetic interactions that ultimately determine the expression levels of genes within the cell. Such interactions are composed into Gene Regulatory Networks (GRNs), which describe how individual genes regulate one another. Computational modeling allows us to represent a mechanistic understanding of GRNs, to formally compare model simulations to experimental data, explore new hypotheses and perform *in-silico* experiments.

Recent findings suggest that the process through which cells take on a specific role, termed differentiation, might be implemented by changing the accessibility of binding sites required for regulation [23], essentially enabling and disabling interactions in the GRN. When considering network reconfiguration in cells, self-modifying programs come to mind. Self-modifying programs are not a new

© Springer International Publishing Switzerland 2015
M. Lones et al. (Eds.): IPCAT 2015, LNCS 9303, pp. 131–144, 2015.
DOI: 10.1007/978-3-319-23108-2_11

concept in software, but they have not become mainstream, mainly because in most contexts they do not add expressive power, and they are hard to write and analyze. Consequently, modern program analysis tools have no, or very limited, means of reasoning about such programs. It does appear, however, that for biological modeling, supporting the concept of switching networks can provide a useful abstraction for capturing the processes at work as cells change type.

To capture these phenomena, we introduce the concept of a *Switching Gene Regulatory Network* (SGRN), a framework for the analysis and synthesis of self-modifying biological programs. An SGRN is constructed to incorporate knowledge of network topology and to reproduce and explain experimental observations of system dynamics, by integrating known biological hypotheses. We formalize our approach and provide an encoding of SGRNs and bounded temporal constraints representing known experimental data, within a framework based on Satisfiability Modulo Theories (SMT) solvers. This builds upon and extends our previous work in the area, which supported only fixed GRNs [8, 26]. Finally, we evaluate the performance of our approach on a set of synthetic benchmarks in terms of running time, accuracy, and precision and we show that our method is scalable and that it reliably recovers the changes taking place in the network topology.

## 2    Background

We focus on *Boolean networks (BNs)* [13], a class of GRN models that are Boolean abstractions of genetic systems, *i.e.* every gene is represented by a Boolean variable specifying whether the gene is active or not. The concept of an *Abstract Boolean Network (ABN)* was introduced in [8] to allow the representation of models with initially unknown network topologies and dynamics. ABNs were then used to investigate the decision-making in pluripotent stem cells. In the following, we briefly review the relevant definitions from [8], which serve as a basis for the modeling approach described in later sections.

Let $G$ be a finite set of genes and let $E : G \times G \times \mathbb{B} \to \mathbb{B}$ denote the set of directed edges between elements of $G$, labeled with a regulation activity ($\top$ for positive and $\bot$ for negative). Given genes $g$ and $g'$, we call $g$ an *activator* of $g'$ if $(g, g', \top) \in E$, a *repressor* if $(g, g', \bot) \in E$ and a *regulator* if it is either of those. Due to the Boolean abstraction of genetic states, the state space $Q = \mathbb{B}^{|G|}$ is induced implicitly where, for a given state $q \in Q$ and gene $g \in G$, $q(g) \in \mathbb{B}$ denotes the state of $g$. An update function $f_g : Q \to \mathbb{B}$ defines the dynamics of gene $g$. For a Boolean network with synchronous updates, the dynamics of the system are defined in terms of the update functions of all genes applied at each step, where given a current and next state $q, q' \in Q$, $\bigwedge_{g \in G} q'(g) = f_g(q)$. Although the presentation and examples in this paper focus on synchronous semantics, we also support asynchronous updates, where at each step the update function of only one gene is applied, while the value of all the other genes remains unchanged.

A set of 18 biologically plausible update function templates, which are called *regulation conditions*, was proposed in [8]. For a given gene $g \in G$, each regulation condition defines an update function $f_g : E \times Q \to \mathbb{B}$ that respects biologically-inspired constraints. One such constraint is monotonicity, where the availability of additional activators does not lead to the inactivation of a gene, i.e., if a gene is expressed in $q'$ when only some of its activators are expressed in $q$, then it must also be expressed in $q'$ if all its activators are expressed in $q$ and there is no change in the presence of repressors. These regulation conditions only consider whether none, some, or all potential activators or repressors of $g$ are expressed in a state $q$.

To capture the possible uncertainty in the precise network topology, the ABN formalism allows some interactions to be marked as *optional* (denoted by the set $E^?$), each of which could be included in a synthesized *concrete model* (a model where all interactions are definite). Thus, in terms of network topology, an ABN model specifies a set of $2^{|E^?|}$ concrete models, each corresponding to a unique selection of optional interactions. Additionally, a choice of several possible regulation conditions for each gene is allowed, leading to the following definition:

**Definition 1 (Abstract Boolean Network [8]).** An abstract Boolean network (ABN) is a tuple $\mathcal{N} = (G, E, E^?, R)$, where $G$ is a finite set of genes, $E : G \times G \times \mathbb{B} \to \mathbb{B}$ is the set of definite (positive and negative) interactions between them, $E^? : G \times G \times \mathbb{B} \to \mathbb{B}$ is the set of optional interactions and $R = \{R_g \mid g \in G\}$, where $R_g$ specifies a (non-empty) set of possible regulation conditions for each gene $g \in G$.

An ABN is transformed into a concrete model by selecting a subset of the optional interactions to be included and assigning a specific regulation condition for each gene. Formally, let $\hat{E}^? \subseteq E^?$ denote the set of selected optional interactions, $\hat{E} = E \cup \hat{E}^?$ denote the set of all selected interactions and $\hat{R}_g \in R_g$ denote the specific regulation condition chosen for each gene $g \in G$. The semantics of such a concrete model are defined in terms of a transition system $\mathcal{T} = (Q, T)$, where $Q = \mathbb{B}^{|G|}$ is the set of states ($q(g) \in \mathbb{B}$ is the state of gene $g$ in $q \in Q$) and the transition relation $T : Q \times Q \to \mathbb{B}$ is defined as

$$\forall q, q' \in Q . \, T(q, q') \leftrightarrow \bigwedge_{g \in G} q'(g) = \hat{R}_g(\hat{E}, q). \tag{1}$$

A finite *trajectory* of a concrete model is defined as a sequence of states $t = q_0, q_1, \ldots, q_K$ where $q_i \in Q$ and $\forall_{0 \le i < K} . \, T(q_i, q_{i+1})$. The semantics of an ABN can be understood in terms of the choice of optional interactions $\hat{E}^?$ and the choice of a regulation condition for each gene, $\hat{R}_g$, together with the transition system $\mathcal{T}$ representing the resulting concrete model.

A set of *experimental observations* that each concrete model needs to be able to satisfy are encoded as predicates over system states which limits the possible consistent choices of $\hat{E}^?$ and $\hat{R}_g$. For instance, an *experiment* in which genes $g$ and $g'$ are observed to be initially active and become inactive at step $K$ is formalized as a constraint requiring the existence of a trajectory $t = q_0, \ldots, q_K$ such that $q_0(g) \wedge q_0(g') \wedge \neg q_K(g) \wedge \neg q_K(g')$. The approach developed in [8] allows GRN synthesis for non-switching networks: given an ABN and a set of experiments, find a choice of interactions $\hat{E}^?$ and regulation conditions $\hat{R}_g$ for each gene, guaranteeing that the resulting concrete model is consistent with all experimental observations.

# 3    Switching Gene Regulatory Networks

We propose an extension of the ABN formalism, where transitions between unique cell types, characterized by potentially different network topologies, are directly supported. Let $C$ denote a set of cell types sharing a set of genes $G$ and regulation conditions $R$. Each cell type $c \in C$ is modeled as an ABN $\mathcal{N}_c = (G, E_c, E_c^?, R)$, where the set of definite interactions $E_c$ and optional interactions $E_c^?$ could be different for different cell types. Note that, while the network topology is allowed to change between different cell types, we assume that the possible regulation conditions $R_g \in R$ depend only on a gene $g \in G$ and remain consistent across cell types.

Arbitrary transitions between different cell types are not plausible in most biological systems. For example, two distinct cell types $c, c' \in C$ can represent a progenitor cell $c$ and a differentiated cell $c'$ that is derived from $c$. While the progenitor can become a differentiated cell, the reverse does not occur under normal conditions. For each cell $c \in C$, we capture this information using the (non-empty) subset $D_c \subseteq C$ of all possible cell types that $c$ can transition into directly. In order to capture mechanistic details within the model, our framework also supports the addition of guards, encoded as state predicates, to further constrain cell type switches. In the absence of restrictive guards, switching between cell types is represented as a nondeterministic choice (when $|D_c| > 1$), without explicitly modeling either the mechanism or preconditions on the system state required for such a switch.

This leads to the following definition of SGRNs:

**Definition 2 (Switching Gene Regulatory Network).** A Switching Gene Regulatory Network (SGRN) is a tuple $\mathcal{N}_S = (G, C, D_c, E_c, E_c^?, R)$, where

- $G$ is the finite set of genes,
- $C$ is a finite set of cell types,
- for each $c \in C$, $D_c \subseteq C$ is the set of cell types that $c$ can transition into directly,
- $E_c : G \times G \times \mathbb{B} \to \mathbb{B}$ is the set of definite interactions between genes for each $c \in C$,
- $E_c^? : G \times G \times \mathbb{B} \to \mathbb{B}$ is the set of optional interactions for cell type $c$, and
- $R = \{R_g \,|\, g \in G\}$, where $R_g$ specifies a (non-empty) set of possible regulation conditions for each gene $g \in G$.

Figure 1 shows an SGRN with 3 cell types: $C = \{c_0, c_1, c_2\}$, and 6 genes: $G = \{g_0, g_1, g_2, g_3, g_4, g_5\}$. In this example, a (progenitor) cell type, $c_0$, may change into cell types $c_1$ or $c_2$, by reconfiguring its network, so that $D_{c_0} = \{c_1, c_2\}$, while

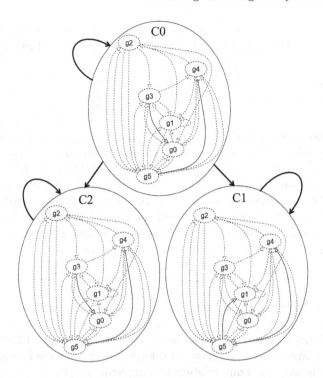

**Fig. 1.** An SGRN with 3 cell types (c0-c2) with 6 genes (g0-g5), illustrating a typical setting where one cell type (c0) can maintain its identity (self-loops) or give rise to other cell types by switching its interactions. Edges between genes represent regulatory interactions, with a bar representing repression and an arrow representing activation, and appear as solid or dashed lines for definite or optional interaction, respectively.

$c_1$ and $c_2$ cannot switch their identity (thus $D_{c_1} = \{c_1\}$ and $D_{c_2} = \{c_2\}$). For each cell type, edges between genes appear in solid or dashed lines for definite ($E_c$) or optional ($E_c^?$) interactions respectively. Genes appear in dashed circles to indicate that $R$ consists of multiple possible regulations conditions for each gene.

As in Sect. 2, the semantics of SGRNs are defined in terms of a transition system $\mathcal{T} = (Q, T)$. Here $Q = \mathbb{B}^{|G|} \times C$ is the set of states where, for a given state $q \in Q$, $q = (q_G, q_C)$, $q_G(g) \in \mathbb{B}$ indicates the state of a given gene $g \in G$ and $q_C \in C$ indicates the current cell type. For a concrete switching GRN model, let $\hat{E}_c^? \subseteq E_c^?$ denote the set of selected optional interactions and $\hat{E} = E_c \cup \hat{E}_c^?$ denote the set of all selected interactions for each cell type $c \in C$. Let $\hat{R}_g \in R_g$ denote the specific regulation condition selected for each gene $g \in G$, which is the same for all cell types. The transition relation $T : Q \times Q \to \mathbb{B}$ is defined as

$$\forall q, q' \in Q.\ T(q, q') \leftrightarrow \bigwedge_{c \in C} \left[ q_C = c \to \left( q'_C \in D_c \land \bigwedge_{g \in G} q'_G(g) = \hat{R}_g(\hat{E}_c, q_G) \right) \right].$$

$$(2)$$

Intuitively, Eq. 2 captures the fact that all genes are updated according to the selected regulation conditions $\hat{R}_g$ and the network topology $\hat{E}_c$ corresponding to the particular cell type $c$ in the current state $q$. In the next state $q'$, the cell type can be updated (non-deterministically) to one of the possible cell types $D_c \subseteq C$ that $c$ can transition into directly. As for ABNs, given an assignment of the optional interactions $\hat{E}_c^?$ for each cell type $c \in C$, and a specific regulation condition $\hat{R}_g$ for each gene $g \in G$, Eq. 2 allows us to define finite trajectories of the resulting concrete SGRN models as a sequence of states $t = q_0, q_1, \ldots, q_K$ from $Q$ where $\forall_{0 \le i = 0 < K} \cdot T(q_i, q_{i+1})$.

## 4   SGRN Model Synthesis

We are interested in concrete SGRN models that are consistent with given experimental observations. In this section, we formalize this as a synthesis problem and present the details of our solution and implementation.

A SGRN model $\mathcal{N}_S = (G, C, D_c, E_c, E_c^?, R)$ is transformed into a concrete model by selecting a specific regulation condition $\hat{R}_g \in R_g$ for each gene $g \in G$ and a subset of the optional interactions $\hat{E}_c^? \subseteq E_c^?$ to be included for each cell type $c \in C$. Each possible concrete model is represented as a transition system $\mathcal{T} = (Q, T)$, where the system set of states is $Q = \mathbb{B}^{|G|} \times C$.

Let $\pi : Q \to \mathbb{B}$ denote a state predicate capturing some observed gene states or cell type and the tuple $(\pi, n)$ denote a constraint that, for a given trajectory $t = q_0, \ldots, q_K$, $t$ satisfies $\pi$ at step $n$ (i.e. $\pi(q_n) = \top$). An *experiment* $\mathcal{E} = \{(\pi_i, n_i) \mid i = 0 \ldots M\}$, where $\pi_i$ is a state predicate and $n_i \in [0, K]$ for all $i \in [0, M]$, is expressed as a finite set of such constraints and formalizes the gene expressions or cell types observed during a particular execution of the system. We write $t \vDash \mathcal{E}$ when trajectory $t$ satisfies experiment $\mathcal{E}$ (i.e. when $\bigwedge_{(\pi,n) \in \mathcal{E}} \pi(q_n)$). More complicated expressions can also be constructed as part of an experiment by combining terms $(\pi, n)$ using the logical operators $\{\land, \lor, \Rightarrow, \Leftrightarrow, \neg\}$.

The main problem we consider in this paper is the following:

**Problem 1 (Lineage Synthesis)** *Given an SGRN $\mathcal{N}_S = (G, C, D_c, E_c, E_c^?, R)$ and a finite set of experiments $\mathcal{E}_0, \ldots, \mathcal{E}_m$, find an assignment $\hat{E}_c^?$ of the optional interactions $E_c^?$ for each cell type $c \in C$ and a single regulation condition $\hat{R}_g \in R_g$ for each gene $g \in G$ such that, for each $i = 0, \ldots, m$ there exists a trajectory $t_i$ of the resulting concrete model that satisfies $\mathcal{E}_i$ (i.e. $t_i \vDash \mathcal{E}_i$).*

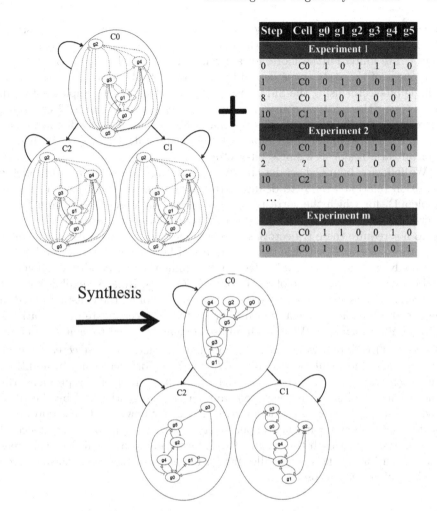

| Step | Cell | g0 | g1 | g2 | g3 | g4 | g5 |
|------|------|----|----|----|----|----|----|
| Experiment 1 | | | | | | | |
| 0 | C0 | 1 | 0 | 1 | 1 | 1 | 0 |
| 1 | C0 | 0 | 1 | 0 | 0 | 1 | 1 |
| 8 | C0 | 1 | 0 | 1 | 0 | 0 | 1 |
| 10 | C1 | 1 | 0 | 1 | 0 | 0 | 1 |
| Experiment 2 | | | | | | | |
| 0 | C0 | 1 | 0 | 0 | 1 | 0 | 0 |
| 2 | ? | 1 | 0 | 1 | 0 | 0 | 1 |
| 10 | C2 | 1 | 0 | 0 | 1 | 0 | 1 |
| ... | | | | | | | |
| Experiment m | | | | | | | |
| 0 | C0 | 1 | 1 | 0 | 0 | 1 | 0 |
| 10 | C0 | 1 | 0 | 1 | 0 | 0 | 1 |

**Fig. 2.** A lineage synthesis problem. The SGRN from Fig. 1 and a finite set of experiments define a lineage synthesis problem. A solution for this problem includes the assignment of definite interactions for each cell type and the choice of a single regulation condition for each gene.

Figure 2 illustrates a lineage synthesis problem for an example SGRN.

Given an SGRN $\mathcal{N}_S = (G, C, Dc, E_c, E_c^?, R)$ we encode the choice of optional interactions $\hat{E}_c^?$ for each cell type $c \in C$ using a unique Boolean choice variable for each interaction, or more conveniently, as a single bit-vector using the respective SMT theory. Additionally, a single regulation condition $\hat{R}_g$ from the set of allowed conditions $R_g$ must be selected for each gene $g \in G$. We encode this as the synthesis of a single bit-vector or integer 'coefficient' for each gene, which is shared across all cell types.

The choice variables for optional interactions of each cell type and regulation conditions for each gene allow us to consider the transition system $\mathcal{T} = (Q, T)$

as defined in Sect. 3, which represents a given concrete model. The set of states $Q = \mathbb{B}^{|G|} \times C$ is finite since both the number of genes $G$ and the number of cell types $C$ are finite. Furthermore, for a given state $q \in Q$ where $q = (q_G, q_C)$, the component of the state space describing the state of all genes $q_G$ is encoded as a single bit-vector using the SMT theory of bit-vectors. In our implementation, we represent the cell type component of a state $q_C$ using a "one-hot" encoding, where $q_C \in \mathbb{B}^{|C|}$ with the guarantee that the cardinality of $q_C$ for any state $q \in Q$ is 1. This allows us to represent the entire state $(q_G, q_C)$ as individual Boolean variables or as a single bit-vector.

We follow a bounded model checking (BMC) approach [4], and unroll the transition relation $T$ of $\mathcal{T}$ to define a trajectory $t_i$ for each experiment $\mathcal{E}_i$ (see problem 1), for which the corresponding experimental observations from $\mathcal{E}_i$ are asserted. Note that while a separate trajectory $t_i$ is used for each experiment $\mathcal{E}_i$, we do not require these trajectories to be unique (*i.e.* it is possible that a single trajectory $t = t_i = t_j$ satisfies the constraints of both experiments $\mathcal{E}_i$ and $\mathcal{E}_j$).

Finally, we employ an SMT solver to determine the satisfiability of all generated constraints (our choice of SMT solver is Z3 [16]). Here, we exploit the fact that SMT solvers such as Z3 produce an assignment of all the constants used in the encoding of the problem, which is presented as a certificate of the satisfiability of all constraints. When such an assignment (referred to as a "model" in this context) is found, we extract the optional interactions $\hat{E}_c^?$ selected for each cell type and the regulation condition $\hat{R}_g$ selected for each gene. In addition, since each trajectory $t_i$ was represented explicitly as part of the problem, the exact sequence of states is recovered from the model synthesized by the SMT solver, to serve as an example demonstrating exactly how the SGRN reproduces the behavior observed in each experiment $\mathcal{E}_i$. In addition to the sequence of gene expressions at each time point, this information also reveals the cell types along executions of the system, allowing for further investigation of the captured cellular differentiation processes.

# 5  Experimental Results

In order to test our approach and systematically evaluate its performance we require benchmarks of lineage synthesis problems for SGRNs with different number of genes and cell types. This is achieved by producing synthetic problems, following the main steps summarized in Fig. 3 and described in Subsect. 5.1. Subsection 5.2 gives the results of our evaluation in terms of accuracy, precision and running time.

## 5.1  Benchmark Design

Cell types are defined by directed networks with a scale-free topology (the degree of the vertices follows a power-law distribution), which is a common feature of GRNs and other biological networks [2], with the exponent of the degree distribution set to 2 (for both in- and out- degree distributions). Interactions are

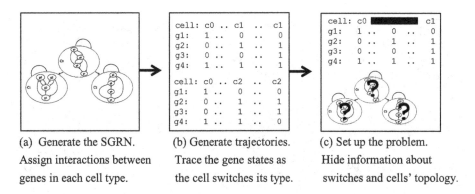

(a) Generate the SGRN.
Assign interactions between
genes in each cell type.

(b) Generate trajectories.
Trace the gene states as
the cell switches its type.

(c) Set up the problem.
Hide information about
switches and cells' topology.

**Fig. 3.** The three steps for generating an *in-silico* lineage synthesis problems involve: (a) randomly generating a concrete SGRN, where all interactions are definite and a single regulation condition is allowed for each gene. (b) Generating trajectories of the concrete SGRN model from (a). This essentially amounts to simulation, which is possible since the model does not include any uncertainty. (c) Generating a lineage synthesis problem with partial information about the interactions in the system (encoded as an SGRN) and the trajectories it produces (encoded as experimental observations).

labelled with either a positive or negative sign, such that each gene has at least one activator. This is in keeping with the assumption that, by default, genes are repressed in higher organisms, and must be "switched on" to be expressed and behave as regulators of their target genes [19]. A regulation condition is randomly assigned to each gene from a set of 16 out of the 18 regulation conditions defined in [8], excluding the two functions that allow activation of a gene in the absence of any activators. For a given model with $m$ cell types, $n$ genes, and a progenitor cell type $c_0$, we generate $2 \cdot m \cdot n$ trajectories of length $K = 11$ starting at $c_0$ with a gene state configuration $j$, and switching to cell type $c_i$ at a randomly selected time point $s$, for $i = 0 .. m$, $j = 1 .. 2n$ and $1 \leq s \leq K$. In order to create the set of $2n$ starting gene state configurations, we randomly select $2n - 2$ integer values in the range $(0, 2^n - 1)$ (exclusive) and add the values 0 and $2^n - 1$, representing the extreme configurations of the system. System states are represented by bit-vectors of size $|G|$, where the $k^{th}$ position in the vector represents the state of the $k^{th}$ gene.

To construct an instance of the lineage synthesis problem, each model (generated as described above) is used to produce a SGRN and its trajectories are encoded as experimental observations. We assume no information about the exact regulation conditions available and, therefore, all 16 choices are allowed for each gene. Let $E_c^\star$ denote the interactions of cell type $c$ in the "true" model and $E^\star = \cup_{c \in C} E_c^\star$ denote the interactions appearing in any cell type. We construct the SGRN by assigning a small proportion (20 %) of $E_c^\star$ as definite for cell type $c$ (representing known interactions) and marking the rest of $E^\star$ as optional, which defines the sets $E_c$ and $E_c^?$ respectively (Fig. 4). Each trajectory is then used to generate an experiment with the gene states observed at each time step, and the cell type observed at the start and at the end of the experiment (time

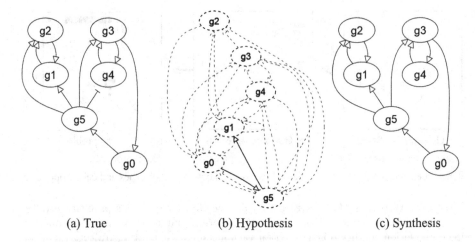

(a) True                    (b) Hypothesis              (c) Synthesis

**Fig. 4.** A true, a hypothesized, and a synthesized progenitor cell type in an SGRN with 6 genes (g0-g5) and 3 cell types. The true cell type (a) was generated with a scale-free topology. The union of all cell types in the SGRN was used to create the hypothesized cell type (optional interactions appear as dashed lines) with a small proportion of its true interactions known (solid lines). Genes appear with dashed circles to indicate that their regulation condition is not fixed. The synthesized cell type is part of our solution for a lineage synthesis problem generated for this SGRN and recovers the true cell type with the exception of the negative interaction from g5 to g4.

steps 0 and 10, correspondingly). In total, this amounts to $2 \cdot m \cdot n$ experiments included in a lineage synthesis problem of $m$ cells and $n$ genes.

## 5.2   Results

We demonstrate our technique on benchmarks of lineage synthesis problems with 1–7 cell types and 4–10 genes, generated as described above. For each problem we record the running time required to solve and we evaluate solutions by means of accuracy and precision in relation to the 'hidden' true model from which each problem was generated.

Let $E_c^\star$ denote the "true" interactions of cell type $c$, $E_c$ ($E_c^?$) denote the definite (optional) interactions of the corresponding SGRN cell type, and $\hat{E}_c$ ($\hat{E}_c^?$) denote the synthesized (optional) interactions. A *True Positive* (*Negative*) is an interaction that is (not) in $\hat{E}_c^?$ and (not) in $E_c^\star$ (note that we evaluate the synthesis of only those interactions that were optional in the SGRN since definite interactions will always be part of the synthesized model). A *False Positive* is an interaction in $\hat{E}_c^?$ that is not in $E_c^\star$ and a *False Negative* is an interaction in $E_c^\star$ that is not in $\hat{E}_c^?$. The precision of a solution for a given cell type is then defined as $\frac{TP}{TP+FP}$, and its accuracy as $\frac{TP+TN}{TP+FP+TN+FN}$, with $TP, TN, FP$ and $FN$, the number of True Positives, True Negatives, False Positives and False Negatives, respectively. The total precision and accuracy of a solution is the mean precision and accuracy across all cell types in the problem.

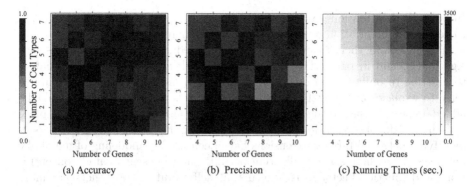

**Fig. 5.** Heatmaps of experimental results for a benchmark of lineage synthesis problems with 1–7 cells and 4–10 genes. Darker pixels indicate higher accuracy (a) and precision (b), while lighter pixels indicate poorer performance. Running times (c) are indicated on a color scale from white to black, with darker pixels for longer running times.

The results of our evaluation (Figs. 5a,b) show that our approach can successfully recover hidden topologies of SGRNs, achieving 0.81 accuracy and 0.78 precision (on average, across 2–7 cell types and 1–10 genes). As evident from the heatmaps in Figs. 5a,b, cell types are synthesized with good accuracy across problems (17 % with accuracy > 0.9, 86 % of cases with accuracy > 0.7 and all problems with accuracy > 0.6) and with good precision in the majority of cases (71 % of cases with precision > 0.7). For our benchmarks, the performance seems to be independent of the number of cells or genes. The running time of our synthesis is also feasible for the SGRNs under consideration, with all problems in the benchmark set solved in under an hour on a personal computer (Intel Core i3-4010U 1.7 GHz, 4 GB RAM, Windows 8.1 64-bit OS) and with an average running time of 730.25 sec (Fig. 5c).

## 6  Related Work

Since the early days of computer science, the concept of self-modifying programs has been a natural one to explore, especially after the introduction of the Von Neumann architecture [17], in which both the program and the data were stored in the same memory, leading to the possibility of allowing program modification during runtime. This model was supported in early computer architectures (cf. e.g., [3]) and applied in some specific domains, however it did not become a mainstream paradigm.

Boolean networks have been suggested for studying cell differentiation [13, 24]. In this context the concept of *switching* was mainly used to describe changes in the state of the nodes (genes) rather than the reconfiguration of the topology of the network itself. The change in the gene's state could be a result of executing the GRN and by including additional effects such as the spatio-temporal dynamics of the neighbouring cellular (tissue) environment (for example: [7,10]).

However, little attention was given to the rewiring of the network as a mechanism to achieve differentiation or changes in the cellular function.

Petrinets and their extensions have been used in modeling of GRNs (see e.g., [5,11]) and in particular the extension of self-modifying nets [25] enables to describe reconfiguration of Petrinets. This is achieved by allowing an arc to refer to a place, implying that the number of tokens in this place should be added/removed while firing the transition. The number of tokens in a place can change during execution leading to the 'reconfiguration' of the net. Self-modifying nets and further extensions have been used in modeling of metabolic networks [12], where self-modification permits the representation of concentrations and kinetic effects. It is known that self-modifying Petrinets are more expressive than conventional Petrinets, making the reachability problem undecidable [25], whereas in our work we defined a framework in which the basic dynamic properties of the system remain decidable.

Bayesian networks have been extensively applied to the problem of inference of gene regulatory networks from time series data [9]. Unlike our work, these methods handle continuous variables and stochastic events, but they lack some of the general advantages of reasoning based approaches, including proofs that solutions do not exist and effective ways to symbolically reason about sets of solutions. More recently, there has been research on generalizing Bayesian networks inference to the case of time varying networks (e.g., [1,6,14,18,21,22]).

Related concepts of *switching* have also been introduced and explored in other fields. For example, mode-automata was proposed as a formalism for modelling reactive systems, in order to capture explicitly a decomposition of the system's global behaviour into multiple independent tasks [15]. In our work, however, such a decomposition is not fully known a priori and our focus is on synthesizing the structure of the system in different cell types, which can be viewed as modes, together with the transitions between them. Thus, our approach is also related to methods for the synthesis of controllers for discrete event systems (e.g. [20]) - a problem that has received considerable attention. However, the problem we address requires the synthesis of a system for each cell type such that the overall behaviour reproduces certain experimental observations, rather than synthesizing a controller that, when coupled with the system, restricts its behaviour to some desirable subset.

# 7   Conclusion

Computational methods are becoming a powerful tool for experimental biologists to improve the understanding of cellular decision-making. In particular, formal reasoning and different synthesis approaches are attractive as they enable the automatic generation of models that are guaranteed to satisfy a given set of constraints representing known experimental measurements. Motivated by recent biological evidence suggesting that it makes sense to view a molecular program within a cell as a self-modifying program, we introduce a framework that allows us to represent cellular reconfiguration, and effectively synthesize models that are consistent with experimental constraints and hypotheses. This opens the way to combined computational and experimental research to improve our understanding of how cells differentiate into specific cell types during development, as well as how cells may modify their behavior under artificial culture conditions used for research and medical applications. A long-term research goal is to gain a mechanistic understanding of how self-modifying biological programs operate and investigate whether the underlying principles nature utilizes can inspire new directions for the design of self-modifying software.

**Acknowledgments..** Yoli Shavit is supported by the Cambridge International Scholarship Scheme (CISS). The research was carried out during her internship at Microsoft Research Cambridge, UK.

# References

1. Ahmed, A., Xing, E.: Recovering time-varying networks of dependencies in social and biological studies. In: Proceedings of the National Academy of Sciences, vol. 106, no. 29, pp. 11878-11883 (2009)
2. Albert, R.: Scale-free networks in cell biology. J. Cell Sci. **118**(21), 4947–4957 (2005)
3. Bashe, C., Johnson, L., Palmer, J., Pugh, E.: IBM's early computers. MIT Press (1986)
4. Biere, A., Cimatti, A., Clarke, E., Zhu, Y.: Symbolic model checking without BDDs. In: Cleaveland, W.R. (ed.) TACAS 1999. LNCS, vol. 1579, p. 193. Springer, Heidelberg (1999)
5. Chaouiya, C.: Petri net modelling of biological networks. Briefings Bioinform. **8**(4), 210–219 (2007)
6. Dondelinger, F., Lébre, S., Husmeier, D.: Non-homogeneous dynamic Bayesian networks with Bayesian regularization for inferring gene regulatory networks with gradually time-varying structure. Mach. Learn. **90**(2), 191–230 (2013)
7. Doursat, R.: The growing canvas of biological development: multiscale pattern generation on an expanding lattice of gene regulatory nets. In: Minai, A., Braha, D., Bar-Yam, Y. (eds.) Unifying Themes in Complex Systems, pp. 205–210. Springer, Heidelberg (2008)
8. Dunn, S., Martello, G., Yordanov, B., Emmott, S., Smith, A.: Defining an essential transcription factor program for naïve pluripotency. Sci. **344**(6188), 1156–1160 (2014)
9. Friedman, N., Linial, M., Nachman, I., Pe'er, D.: Using Bayesian networks to analyze expression data. J. Comp. Bio. **3**(7), 601–620 (2000)
10. Giavittob, J., Klaudela, H., Pommereau, F.: Integrated regulatory networks (IRNs): spatially organized biochemical modules. Theoret. Comput. Sci. **431**, 219–234 (2012)
11. Heiner, M., Gilbert, D., Donaldson, R.: Petri nets for systems and synthetic biology. In: Bernardo, M., Degano, P., Zavattaro, G. (eds.) SFM 2008. LNCS, vol. 5016, pp. 215–264. Springer, Heidelberg (2008). Advanced Lectures
12. Hofestädt, R., Thelen, S.: Quantitative modeling of biochemical networks. Silico Biol. **1**(1), 39–53 (1998)

13. Kauffman, S.: Metabolic stability and epigenesis in randomly constructed genetic nets. J. Theor. Biol. **22**(3), 437–467 (1969)
14. Khan, J., Bouaynaya, N., Fathallah-Shaykh, H.: Tracking of time-varying genomic regulatory networks with a LASSO-Kalman smoother. EURASIP J. Bioinf. Sys. Bio. **1**(2014), 1–15 (2014)
15. Maraninchi, F., Rémond, Y.: Mode-automata: about modes and states for reactive systems. In: Hankin, C. (ed.) ESOP 1998. LNCS, vol. 1381, p. 185. Springer, Heidelberg (1998)
16. de Moura, L., Bjørner, N.S.: Z3: An Efficient SMT Solver. In: Ramakrishnan, C.R., Rehof, J. (eds.) TACAS 2008. LNCS, vol. 4963, pp. 337–340. Springer, Heidelberg (2008)
17. von Neumann, J.: First draft of a report on the EDVAC. Technical report Contract No. W670ORD4926, Moore School of Electrical Engineering, University of Pennsylvania (1945)
18. Parikh, A., Wu, W., Curtis, R., Xing, E.: TREEGL: reverse engineering tree-evolving gene networks underlying developing biological lineages. Bioinf. **27**(13), i196–i204 (2011)
19. Phillips, T.: Regulation of transcription and gene expression in eukaryotes. Nat. Educ. **1**(1), 199 (2008)
20. Ramadge, P.J., Wonham, W.M.: Supervisory control of a class of discrete event processes. SIAM J. Control Optim. **25**(1), 206–230 (1987)
21. Rao, A., Hero, A., States, D., Engel, J.: Inferring time-varying network topologies from gene expression data. EURASIP J. Bioinformatics Syst. Biol. **2007**, 7–7 (2007)
22. Song, L., Kolar, M., Xing, E.: Time-varying dynamic Bayesian networks. In: Advances in Neural Information Processing Systems (NIPS) pp. 1732-1740 (2009)
23. Stergachis, A.B., et al.: Developmental fate and cellular maturity encoded in human regulatory DNA landscapes. Cell **154**(4), 888–903 (2013)
24. Thomas, R., Kaufman, M.: Multistationarity, the basis of cell differentiation and memory. ii. logical analysis of regulatory networks in terms of feedback circuits. Chaos: An Interdisciplinary J. Nonlinear Sci. **11**(1), 180–195 (2001)
25. Valk, R.: Self-modifying nets, a natural extension of Petri nets. In: Ausiello, G., Böhm, R. (eds.) Colloquium on Automata, Languages and Programming. LNCS, pp. 464–476. Springer, Heidelberg (1978)
26. Yordanov, B., Wintersteiger, C., Hamadi, Y., Kugler, H.: Z34Bio: an SMT-based framework for analyzing biological computation. In: SMT (2013)

# The Role of Ago2 in microRNA Biogenesis: An Investigation of miR-21

Gary B. Fogel[1], Ana D. Lopez[2], Zoya Kai[2], and Charles C. King[2(✉)]

[1] Natural Selection Inc., San Diego, CA, USA
[2] Pediatric Diabetes Research Center, University of California,
San Diego, La Jolla, CA 92121, USA
chking@ucsd.edu

**Abstract.** Research into the biology of microRNAs (miRNA) continues to expand rapidly. As a result, their abundance and importance in cellular regulation and disease states, also continues to grow and they are considered master regulators. Despite this greater understanding, key mechanisms regulating global miRNA transcription have remained elusive. This paper addresses a critical issue regarding regulation of miRNA expression. Here, we describe and biochemically characterize a universal regulatory complex that directly binds miRNA genetic loci and regulates transcription of miRNA genes. In addition, our preliminary results provide evidence that miRNA-induced Ago2 binding can result in positive post-transcriptional regulation of many important primary miRNAs. Using chromatin immuno-precipitation (ChIP) assays, our results demonstrate that the human miRNA binding protein Argonaute 2 (Ago2) associates with endogenous promoter DNA from each of the important human miRNA genes investigated to date. Additionally, our data shows a robust, direct interaction between mature miR-21 directed Ago2 and a miR-21 promoter DNA sequence.

## 1 Introduction

The field of miRNA research has grown exponentially since the first report of human miRNAs in 2000. Our understanding of miRNA function, as well as their abundance and importance in cellular regulation and disease states, has also grown. It is estimated that miRNAs regulate more than half of the transcriptome, and they are master regulators of human embryonic stem cell (hESC) differentiation. Although there has been a virtual explosion of publications and tools generated in the miRNA field, key mechanisms regulating global miRNA transcription have remained elusive.

One issue currently facing miRNA researchers is that all components of the protein machinery required for transcription of miRNA genes have not been completely elucidated. Downstream of transcription, however, global processing pathways common to all miRNAs have been published (Finnegan and Pasquinelli 2013). After transcription by PolII, nascent primary miRNA (pri-miRNA) transcripts are processed by two different RNAse III enzymes: Drosha, which generates precursor miRNAs; followed by Dicer, which releases mature miRNAs. Once mature miRNAs bind to

© Springer International Publishing Switzerland 2015
M. Lones et al. (Eds.): IPCAT 2015, LNCS 9303, pp. 145–152, 2015.
DOI: 10.1007/978-3-319-23108-2_12

Argonaute proteins (Ago2 in humans), this complex is dogmatically called a miRNA induced silencing complex (miRISC). Within this complex, mature miRNAs act as guide strands for miRISCs binding directly to target RNAs which, in turn, initiate downstream regulatory events (Finnegan and Pasquinelli 2013).

Though mature miRNA function has become a major focus of academic and commercial research, two critical issues remain unresolved in the relationship of miRNAs to cancer and other diseases. Firstly, the research community lacks a global component for transcriptional regulation of miRNA genes. In 2012, data was published showing that *let-7* miRNA guided Argonaute proteins directly bind and *positively* regulate maturation of pri-*let-7* transcripts in a novel auto-regulatory feedback loop that is conserved across species (Zisoulis et al. 2012) this was the first report of Ago2 and miRNAs targeting a primary miRNA transcript, the first report of miRNAs and Argonaute proteins mediating *positive* regulation of biogenesis, and the first report of a miRNA induced auto-regulatory model of miRNA maturation. However, this model was restricted to *C. elegans let-7* miRNA. In this paper, we present preliminary results establishing that human Argonaute (Ago2) also binds primary transcripts of *several* important human miRNAs, including miR-21, *let-7i*, miR-302, miR-375, and the mir-17-92 cluster. Because of the conservation of sequence and function the *let-7* locus across species and miRNAs, we expect that this is a global mechanism of Ago2 and miRNA induced *positive* auto-regulation of primary miRNA biogenesis in human cells. This has great importance in improving our understanding of miRNA function and information processing through regulation.

Secondly, miRNAs are viewed predominately (or entirely) by the scientific community, as repressors. In fact, the name RNA induced silencing complex (RISC) was created for RNA interference (RNAi) regulation. Though binding of Argonaute family proteins are required for the function of both miRNAs and small interfering RNAs (siRNAs or RNAi), the ~22 bp siRNAs are perfectly complementary when they base pair to their target and, in every reported case, result in silencing or degradation (Ketting 2011).

## 1.1   Ago2

Human Ago2 protein was first named eukaryotic initiation factor 2C 2 (EIF2C2). Soon after, it was accepted that Ago2 bound the 3' end of mRNA targets to arrest translation. We now know that Ago2 binding to mRNA targets can be promiscuous in the 3' and 5' ends (Lee et al. 2009), and our understanding of the roles for Ago2 in small RNA regulation have also evolved to include targeting and biogenesis of noncoding RNAs (ncRNAs) from precursor to mature without Dicer (Yang and Lai 2010), as well as methylation and transcriptional regulation of DNA, and alternative splicing of nascent mRNA transcripts. In short, we now understand that Ago2 is a major player in global regulation on miRNA expression, biogenesis and function. It is the direct interface between the world of small RNAs and their diverse regulatory outcomes (Fig. 1).

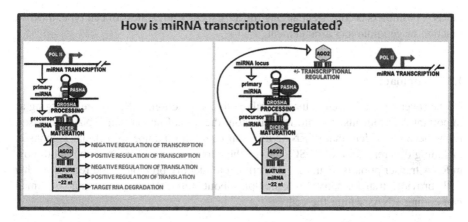

**Fig. 1.** Regulation of miRNA transcription. During miRNA biogenesis, Pol II transcribes the nacent primary miRNA transcript. This pri-miRNA is processed by Drosha which removes the precursor hairpin containing the mature miRNA. This pre-miRNA is further processed by Dicer to generate a mature miRNA that binds Argonaute 2 (Ago2) resulting in multiple regulatory outcomes (left panel). Here we present evidence for a novel miRNA/Ago2 feedback mechansim that regulates miRNA transcription (right panel).

## 1.2    miR-21

Our research initially focused on mature miR-21 because it is one of the most highly expressed human miRNAs, and its importance in cellular regulation and as a potent onco-miR has been the focus of several publications, providing ample information on which to build our experimental design. Additionally, mature miR-21 miRNAs have also been pursued as diagnostic markers for different types of cancer, and pursued as potential therapeutic targets (Kadera et al. 2013; Hong et al. 2013; Jardin and Figeac 2013; Baer et al. 2013). The miR-21 locus is located within the intronic region of the *TMEM49* gene (in the same orientation) and yet transcription of miR-21 is regulated by an independent promoter and terminated by an independent poly (A) tail (Kumarswamy et al. 2011). The miR-21 locus contains a single mature miRNA precursor, as opposed to a cluster of miRNAs, and produces a single primary transcript (Kumarswamy et al. 2011). Transcription of the miR-21 locus is regulated by several important transcription factors, including NFκB and the STAT family of transcriptional activators (Niu et al. 2012). After maturation of the primary miR-21 transcript, a two-step biogenesis process produces the mature miR-21 miRNA. It only after this biogenesis that mature miR-21 can target RNAs for downstream regulation, such as PTEN and PDCD4, which ultimately results in apoptosis and decreased metastasis (Niu et al. 2012). Because mature miR-21 also regulates PDCD4, miR-21 is also capable of preventing type 1 diabetes (T1D) by blocking pancreatic β cell death (Ruan et al. 2011). However, as much as we know about regulation of the miR-21 gene, mechanisms driving this regulation have yet to be identified. Our preliminary data shows that there is at least one additional layer of transcriptional regulation concerning the miR-21, *let-7i*, miR-302, miR-375, and mir-17-92 loci; Ago2 and miRNA induce regulation of transcription, as Ago2 binding to miRNA promoters is consistent among miRNA species, regardless of cell type, miRNA function or genomic loci arrangement.

### 1.3   Summary

Of the several genetic factors that contribute to cell function, miRNAs are emerging as important determinants. As information about the role of non-coding RNAs increases, it has become evident that miRNAs drive an increasing number of cellular responses, including differentiation of hESCs. Given the current dearth of information about how miRNA transcription is regulated, it is critical to identify promoters and repressors that will provide transformative information about information processing of small non-coding RNAs within the cell.

## 2   Methods

CyT49 cells (provided by ViaCyte, San Diego, CA) were maintained on reduced growth factor BD Matrigel at 37 °C, 5 % $CO_2$ in DMEM/F12 supplemented with 20 % knockout serum replacement, glutamax, nonessential amino acids, β-mercaptoethanol, penicillin/streptomycin (Life Technologies, Carlsbad, CA), 4 ng/mL basic fibroblast growth factor (FGF; Peprotech, Rocky Hill, NJ) and 10 ng/mL activin A (R&D Systems, Minneapolis, MN). Cross-linking immunoprecipitation (CLIP) of Ago2, in hESC lines was be used to pull out endogenous DNA. Bound DNA was examined using PCR amplification and sequencing (https://www.idtdna.com).

## 3   Results

Our initial results using cross-linked chromatin immuno-precipitation (ChIP) in human embryonic stem cells (hESCs), and PCR amplification of purified DNA, identified endogenous genes bound by Ago2 (Fig. 2). Using nested primers to increase sensitivity and specificity, we determined that Ago2 binds the promoter of miR-21. Lane 1 is the specific Ago2 (A2) immunoprecipitation (IP), Lane 2 is the negative IP control (using the non-specific HA epitope), Lanes 3 and 4 are positive control inputs for A2 and HA respectively, Lane 5 is the whole cell lysate (W) before IP, and Lane 6 is a negative control with water ($H_2O$) rather than DNA.

In an effort to identify miRNA-binding sequences in the promoter of the miR-21 gene, an RNA hybrid (Kruger and Rehmsmeier 2006) analysis was employed to locate the most likely miR-21 binding region (Fig. 2). After PCR amplification and purification of the ~250 bp region centered around this predicted miR-21 binding site, gel-shift assays with purified Ago2 and synthetic miR-21 RNA oligos were performed. Together, these results demonstrate that Ago2 specifically binds the promoter region of miR-21. Additionally, we have preliminary data suggesting that Ago2 specifically binds the promoter of *let-7i*, miR-375, the miR-302, miR-371, and miR17-92 clusters (manuscript in preparation).

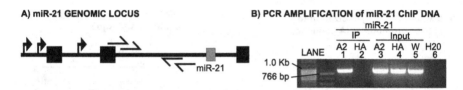

**Fig. 2.** Ago2 binds different miRNA promoter transcripts in human cells. (A) miR-21 genomic locus. The miR-21 precursor (grey box) is between two exons (black boxes) in the same direction of a protein-coding gene. Three start sites (triangle arrows) and four primers (check arrows) are used in nested PCR reactions. (B) PCR amplification of miR-21 ChIP DNA. Lanes 1 and 2 are specific (A2) and non-specific (HA) immunoprecipitations of Ago2, respectively. Lanes 3-5 are positive control input DNA from before each immunoprecipitation (A2 and HA), as well as whole cell lysate (W). Lane 6 is a negative control. All ChIP data shown are representative examples of at least 3 biological replicate experiments.

The data in Fig. 3 shows, that with the addition of the miR-21 molecule and Ago2, the ~250 bp DNA segment is significantly shifted up above 1000 bp with no detectable DNA remaining at ~250 bp. This data supports our ChIP data showing Ago2 targeting of miR-21 promoter DNA, and establishes the interaction as direct and robust.

## 4 Discussion

Taken together, our data demonstrate the importance of mature miR-21 and Ago2 in the binding of the promoter of the miR-21 gene. We hypothesize that this interaction plays an important role driving the transcription of miR-21 biogenesis and ultimately regulates cellular miR-21 levels. Previously, (Zisoulis et al. 2012) found that the Ago2 protein homolog in *Caenorhabditis elegans* specifically binds at the 3' end of *let-7* miRNA primary transcripts to promote downstream processing events in the nucleus and creates a positive-feedback loop. The *C. elegans* studies revealed a novel role for Ago2 in promoting biogenesis of a targeted transcript and autoregulation of *let-7* biogenesis that established a new mechanism for controlling miRNA expression. Here, we extend these initial observations to demonstrate that pri-miRNA-Ago2 interactions occur for multiple mammalian miRNAs and likely serve as previously unrecognized regulators of miRNA transcription.

The canonical view of miRNAs is that these regulatory non-coding RNAs block translation and degrade mRNA. Recently, additional roles of miRNAs in cell translation and regulation of transcription have emerged; suggesting miRNAs act as rheostats to help cells sense their environment and fine-tune the response. Our work has further expanded the role of miRNAs by demonstrating that miRNAs can bind within their own promoters and regulate auto-regulate biogenesis. These observations have the potential to profoundly impact cell biology and modeling progression from normal to disease states. On a cellular level, the mechanisms by which transcription of miRNAs is not well understood. Our biochemical studies indicate that pri-miRNAs bind specific sequences within the miRNA promoters and act as regulators (enhancers or repressors)

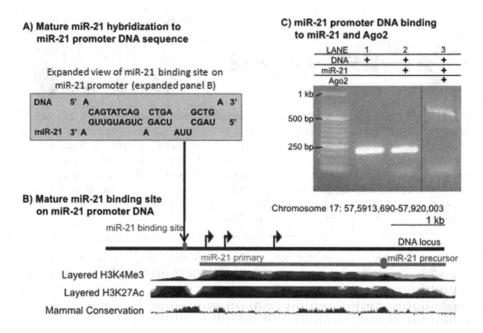

**Fig. 3.** Ago2 directly binds miR-21 promoter DNA. (A) miR-21 locus and mature miR-21 binding. GREY BOX: The top line is the DNA binding sequence; bottom line is the mature miR–21 miRNA sequence. Spaces in sequence represent bulges in miRNA binding. (B) miR-21 locus. BLACK LINE: The human miR-21 locus is on chromosome 17. ARROWS: annotated start sites. SMALLGREY BOX: predicted miR–21 binding site. GREY LINE: nascent primary miR-21 primary transcript. RED BOX: precursor miR-21. 1–3 were taken from the UCSC Genome Browser. (1) Repression of this locus in diverse cell lines as shown by layered Histone 3, lysine 4 methylation (H3K4Me3). (2) Activation of this locus as shown by Histone 3 lysine 27 acetylation (H3K27Ac). (3) Conservation of the DNA sequence across several mammalian species as represented by the line of hash marks represents. (C) Mature miR-21 and Ago2 bind the miR-21 promoter. 1 μg of DNA added in all lanes, 2.5 mM miR-21 synthetic RNA oligo added to lanes 2 and 3. Purified Ago2 (0.03 μg) added to lane 3. Note: bands at ~80 bp in lanes 2&3 are miR-21 primer dimers. Data is representative of five replicates (Color figure online).

of miRNA transcription. Given that binding sequences within individual promoters are at least partially complimentary to the miRNA itself, there exists an opportunity to develop predictive algorithms that can identify binding sites for specific miRNAs. Once identified, interaction sites on the promoter can be tested to determine whether miRNA transcription is enhanced or repressed. Additionally, single nucleotide polymorphism analysis in diseased states can be explored to detect mutations that my alter miRNA binding to specific promoters.

We also note that shortly after the discovery of miRNAs, computational approaches have been used to help elucidate their role in systems biology. However the vast majority of these previous papers have focused solely on miRNAs as drivers of repression (Nissan and Parker 2008; Hobert 2008; Papadopoulos et al. 2009; Djuranovic et al. 2011). While that does not necessarily invalidate their results, it does suggest that alternative computational approaches that allow for both miRNA-induced repression and promotion would provide a more complete view of the cellular mechanics. Further, the true percentage of miRNAs that are involved with promotion remains unclear – it could be that this is a phenomenon only used by a select group of miRNAs or it could be that every miRNA is equally a repressor or promoter depending on the cellular state and environment. The results presented here are focused upon miR-21 because it is known to be involved with many disease pathways including colorectal cancer (Asangani et al. 2008), breast cancer (Frankel et al. 2008) and is already known to be associated as a signature for mycoardial infaraction (Roy et al. 2009) and other human disease issues (Hinton et al. 2012), understanding both its ability to repress and promote is key to understanding its proper function. However, we have idenitifed many other miRNAs that bind to their promoter region, indicating that this regulation is likely to exist for other miRNAs. From this work, a critical question that emerges is: what cellular environments help define the role miRNAs play as either repressors or promoters if it is that they can play both roles? The opportunity to study miRNAs in both roles over multiple environments and cell types will help provide additional understanding and importance to dynamic computational models that can help decipher the meta-level regulatory mechanisms of miRNA regulation. Given the complex dynamic nature of these interactions, methods of computational intelligence and machine learning will be of central importance.

**Acknowledgements.** Funding for these studies was provided by the Larry L. Hillblom Foundation (CCK) and the California Institute for Regenerative Medicine (CIRM).

# References

Finnegan, E.F., Pasquinelli, A.E.: MicroRNA biogenesis: regulating the regulators. Crit. Rev. Biochem. Mol. Biol. **48**(1), 51–68 (2013)

Zisoulis, D.G., Kai, Z.S., Chang, R.K., Pasquinelli, A.E.: Autoregulation of microRNA biogenesis by *let-7* and Argonaute. Nature **486**(7404), 541–544 (2012)

Ketting, R.F.: The many faces of RNAi. Dev. Cell **20**(2), 148–161 (2011)

Lee, I., Ajay, S.S., Yook, J.I., et al.: New class of microRNA targets containing simultaneous 5'-UTR and 3'-UTR interaction sites. Genome Res. **19**(7), 1175–1183 (2009)

Yang, J.S., Lai, E.C.: Dicer-independent, Ago2-mediated microRNA biogenesis in vertebrates. Cell Cycle **9**(22), 4455–4460 (2010)

Kadera, B.E., Li, L., Toste, P.A., et al.: MicroRNA-21 in pancreatic ductal adenocarcinoma tumor-associated fibroblasts promotes metastasis. PLoS ONE **8**(8), e71978 (2013)

Hong, L., Han, Y., Zhang, Y., et al.: MicroRNA-21: a therapeutic target for reversing drug resistance in cancer. Expert Opin. Ther. Targets **17**(9), 1073–1080 (2013)

Jardin, F., Figeac, M.: MicroRNAs in lymphoma, from diagnosis to targeted therapy. Curr. Opin. Oncol. **25**(5), 480–486 (2013)

Baer, C., Claus, R., Plass, C.: Genome-wide epigenetic regulation of miRNAs in cancer. Cancer Res. **73**(2), 473–477 (2013)

Kumarswamy, R., Volkmann, I., Thum, T.: Regulation and function of miRNA-21 in health and disease. RNA Biol. **8**(5), 706–713 (2011)

Niu, J., Shi, Y., Tan, G., et al.: DNA damage induces NF-kappaB-dependent microRNA-21 up-regulation and promotes breast cancer cell invasion. J. Biol. Chem. **287**(26), 21783–21795 (2012)

Ruan, Q., Wang, T., Kameswaran, V., et al.: The microRNA-21-PDCD4 axis prevents type 1 diabetes by blocking pancreatic beta cell death. Proc. Natl. Acad. Sci. USA **108**(29), 12030–12035 (2011)

Kruger, J., Rehmsmeier, M.: RNA hybrid: microRNA target prediction easy, fast and flexible. Nucleic Acids Res. **34**, W451–W454 (2006). (Web Server issue)

Nissan, T., Parker, R.: Computational analysis of miRNA-mediated repression of translation: implications for models of translation initiation inhibition. RNA **14**(8), 1480–1491 (2008). doi:10.1261/rna.1072808

Djuranovic, S., Nahvi, A., Green, R.: A parsimonious model for gene regulation by miRNAs. Science **331**(6017), 550–553 (2011)

Hobert, O.: Gene regulation by transcription factors and microRNAs. Science **319**(5871), 1785–1786 (2008)

Papadopoulos, G.L., et al.: DIANA-mirPath: integrating human and mouse microRNAs in pathways. Bioinformatics **25**(15), 1991–1993 (2009)

Asangani, I.A., et al.: MicroRNA-21 (miR-21) post-transcriptionally downregulates tumor suppressor Pdcd4 and stimulates invasion, intravasation and metastasis in colorectal cancer. Oncogene **27**(15), 2128–2136 (2008)

Frankel, L.B., et al.: Programmed cell death 4 (PDCD4) is an important functional target of the microRNA miR-21 in breast cancer cells. J. Biol. Chem. **283**(2), 1026–1033 (2008)

Roy, S., et al.: MicroRNA expression in response to murine myocardial infarction: miR-21 regulates fibroblast metalloprotease-2 via phosphatase and tensin homologue. Cardiovasc. Res. **82**, 21–29 (2009)

Hinton, A., Hunter, S., Reyes, G., Fogel, G.B., King, C.C.: From pluripotency to islets: miRNAs as critical regulators of human cellular differentiation. Adv. Genet. **79**, 1–34 (2012)

# Evolving Efficient Solutions to Complex Problems Using the Artificial Epigenetic Network

Alexander P. Turner[1](✉), Martin A. Trefzer[1], Michael A. Lones[2], and Andy M. Tyrrell[1]

[1] Department of Electronics, University of York, Heslington, York YO10 5DD, UK
alexander.turner@york.ac.uk
[2] School of Mathematical and Computer Sciences, Heriot-Watt University, Edinburgh EH14 4AS, Scotland, UK

**Abstract.** The artificial epigenetic network (AEN) is a computational model which is able to topologically modify its structure according to environmental stimulus. This approach is inspired by the functionality of epigenetics in nature, specifically, processes such as chromatin modifications which are able to dynamically modify the topology of gene regulatory networks. The AEN has previously been shown to perform well when applied to tasks which require a range of dynamical behaviors to be solved optimally. In addition, it has been shown that pruning of the AEN to remove non-functional elements can result in highly compact solutions to complex dynamical tasks. In this work, a method has been developed which provides the AEN with the ability to self prune throughout the optimisation process, whilst maintaining functionality. To test this hypothesis, the AEN is applied to a range of dynamical tasks and the most optimal solutions are analysed in terms of function and structure.

## 1 Introduction

Biological systems have many innate advantages when compared to typical computational systems. Principally these advantages revolve around the ideas of adaptability, robustness and evolvability [7,10]. It has long since been the goal of computer scientists and engineers to model biological systems in an attempt to capture these properties *in silico*.

Since the inception of biologically inspired computer science, there have been many successes both in capturing the properties of biological systems and using these models to perform computation. It is now commonplace for bio-inspired models to be the state of the art in their field [6,12,19,20]. Certain biologically inspired models are derived from the idea of trying to model their biological counterpart with high levels of detail whereas others take a much more simplistic approach, omitting a vast amount of biological detail. Yet even these simplistic models remain functional and are capable of capturing real-world complex

© Springer International Publishing Switzerland 2015
M. Lones et al. (Eds.): IPCAT 2015, LNCS 9303, pp. 153–165, 2015.
DOI: 10.1007/978-3-319-23108-2_13

**Fig. 1.** DNA being wound round histone octamers into a chromatin fiber

dynamics [2,18]. This poses the question, what is the correct level of abstraction to model a biological system for computation whilst maximising function and reducing model complexity and overheads?

In previous work we have demonstrated that there are significant benefits in increasing the biological faithfulness of artificial gene regulatory networks via the incorporation of an epigenetic analogue. This model, the Artificial Epigenetic Network showed a greater ability to abruptly change its network dynamics resulting in better performance when applied to complex dynamical tasks. In addition the AEN also allowed for the autonomous decomposition of complex tasks into smaller sub-tasks [16,17].

In this work we adapt the networks to allow a more unconstrained type of optimisation which is not limited by the number of inputs and outputs set by a task. This gives the networks potential to significantly alter their size and topology during optimisation to create very small, highly efficient solutions to real world complex tasks.

## 2    Background

Epigenetics refers to mechanisms which result in changes in gene expression without altering the underlying DNA [1]. From both a logical and physical perspective, epigenetics can be considered to be acting on a different level of abstraction compared to genes. A gene can be considered to be a section of DNA which is typically used as an encoding for the primary structure of a protein [11]. Proteins are molecular machines which are responsible for a significant amount of the biochemical interactions within living organisms.

Eukaryotes posses a higher order genetic structure called chromatin. Chromatin acts as a genetic packaging which facilitates the condensing of 2 m of DNA into a 2 μm diameter nucleus within a typical human cell. This is achieved by wrapping the DNA through 1.67 toroidal super helical turns around a histone octamer. This combination of histones and DNA is called the nucleosome and is depicted in Fig. 1. These histone octamers can dynamically change their position on the DNA strand allowing the cellular machinery to access certain genes, effectively acting as a genetic switch. This is the fundamental principle on which the AEN was built upon.

When optimising networks for a given task, it is common practice to generate heuristics about the task and heuristics about the mapping of the task on to the network. For example, if a system contains 10 state variables and is controlled using 2 inputs, it would be fair to assume that the network applied to controlling this system would require at least 10 inputs and produce 2 outputs, omitting any processing units in between. This creates 2 potential knock on effects. Firstly it is assumed that all 10 variables are required in order to solve the task. Secondly, the topology and size of the network are limited according to the task constraints. This is especially problematic when computational efficiency is of importance.

In previous work it has been shown that by comprehensive post processing of the AEN after optimisation, typically involving a brute force search, it is frequently the case that a significant number of genes and epigenetic switches contained in the network are surplus to the network in terms of functionality [15–17]. These surplus units can be removed without altering the performance of the networks, resulting in very small networks which maintain the behaviour of their larger counterparts. This is beneficial because it provides significant information about the exact functionality of the network, the emergence of its behavior and the system it is controlling [15–17].

In the following section, a set of adaptions to the AEN are described which allow for a higher level of plasticity during optimisation by not constraining its size and topology to the dimensions of the task which it is trying to solve. This is done specifically with the idea of the networks being able to create as efficient solution as possible autonomously, during optimisation, without the requirement of significant levels of post processing.

## 3    The Artificial Epigenetic Network

The AEN is built upon an artificial gene regulatory [9] network with the addition of an epigenetic control layer which dynamically alters the expression of genes during execution. It consists of artificial genes and artificial epigenetic switches which are separate in form, but interact functionally during execution of the network.

### 3.1    Artificial Genes

Each gene within the network contains a set of parameters which are listed in Table 1. The inner workings of each gene are a parameterisable sigmoid function. The expression level is calculated using the sigmoid function in Eq. (1), where $s$ (sigmoid slope) $\in [0,20]$, $b$ (sigmoid offset) $\in [-1,1]$ and $x$ is the weighted sum of the expressions of connected genes where $i$ and $j$ are each genes respective weight and expression level (2).

$$f(n) = (1 + e^{-sx-b})^{-1} \tag{1}$$

$$x = \sum_{J=0}^{n} i_j w_j \tag{2}$$

## 3.2   Artificial Epigenetics

The artificial epigenetic control layer is a set of switches (approximately 20 % of the number of genes, which was found to be suitable after prior experimentation [15,16]). The parameters of which can be seen in Table 1. The switches are connected to gene(s) and function according to the same equations the genes use (Eqs. 1 and 2). The difference being that when the result of Eq. 2 is above 0.5, the genes expression values which are connected to the epigenetic molecule are set to 0 (effectively removing them from the network). These values have

**Table 1.** Ranges of the variables within a gene and epigenetic switch. The values are identical to those in [15,16]

| Gene | | |
|---|---|---|
| Variable | Type | Range |
| Gene Expression | Real | 0;1 |
| Weight | Real | -1;1 |
| Sigmoid Offset | Real | -1;1 |
| Sigmoid Slope | Int | 0;20 |
| Input Number | Real | 0;1 |
| Output Number | Real | 0;1 |
| Identification | Real | 0;1 |
| Proximity | Real | 0;0.25 |

| Epigenetic Switch | | |
|---|---|---|
| Variable | Type | Range |
| Identification | Real | 0;1 |
| Proximity | Real | 0;0.15 |
| Sigmoid Offset | Real | -1;1 |
| Sigmoid Slope | Int | 0;20 |

## 3.3   Adaptations

Connections within the network are a product of an $n$-dimensional space where the connections are not directly encoded, but are a product of the genes' interactions within this space and are compiled at runtime. The use of this technique, referred to as indirect representation, has demonstrated advantages in terms of the evolvability of the underlying structure [8,14]. In previous instances an indirect representation was used to derive the connections between genes and epigenetic switches. In this work, this principle is extended to include the mapping of the inputs and outputs of the task onto and from the network.

The implementation of indirect representation in this instance uses 4 variables within each gene. These are the input number, output number, identifier and proximity (Table 1). The input and output number describe a position in two separate one dimensional spaces which specify which environmental input and output, if any, they will be connected to (Fig. 2). Within this space, there are partitions each of which correspond to an environmental input or output. The size of these partitions is created at runtime and equate to 1/average number of genes over all networks. A partition is then attributed to each input, allocating approximately half of the input space to mapping to a specific input. An example

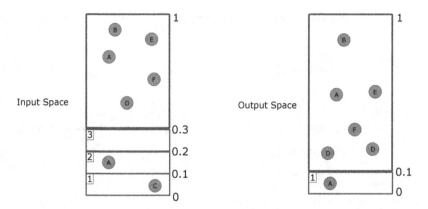

**Fig. 2.** An example of how the inputs and outputs are mapped from the task onto the AEN. Each gene exists within both the input and output space, but only genes within the partitions will be allocated an input or provide an output. This image shows that gene A is connected to input 2, and gene C is connected to input 1. Gene A provides the output for the network

of this can be seen in Fig. 2, where a large proportion of the input space does not map any input onto the gene.

Each gene has a location in the input space, output space and connection space meaning a gene can be any combination of an input, an output, or a processing gene. Upon network initialisation at the start of an experiment, each network is provided with all possible environmental inputs and outputs exactly once. This is achieved by a single gene existing in each partition of the input and output space. The networks are then free to optimise their topology and remove or add additional inputs and outputs from the network during optimisation.

The connections between genes and epigenetic switches use a different space called the connection space which is illustrated in Fig. 3. The connection space is different from the input and output space because genes' connections are a product of both their position and size. Each gene contains an identifier and a size, which specifies a position for that gene within the connection space. Genes and epigenetic switches are connected when their representations in the connection space overlap. This can be seen in Fig. 3 as can the resulting network. In addition, this method also allows for non functioning genes to be effectively bred out of the network by moving to an inactive part of the connection space.

Using this method, environmental inputs are not directly mapped onto a gene. Therefore, the size of the network is no longer constrained to the sum of the environmental inputs and outputs. Only if genes exist within specific places within the input and output spaces will they be allocated inputs or provide outputs.

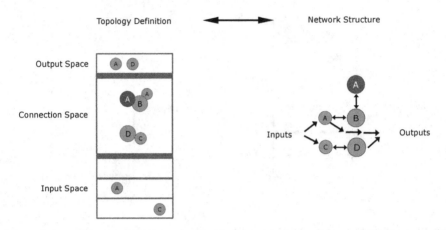

**Fig. 3.** An example of how the input, output and connection space come together to form the topology of the AEN. The genes, light shaded circles exist in the input space, connection space and the output space. Whereas the epigenetic switch A (darker shaded circle) only exists within the connection space. However, because all genes exist within this space also, the epigenetic switch can via proxy also control the inputs and outputs of the AEN

## 4   Methodology

This work sets out to assess if the adaptions made to the AEN allow for the autonomous reduction of network size throughout optimisation whilst maintaining underlying functionality.

To address these questions, the AEN is applied to three tasks. These are the coupled inverted pendulum tasks [5], a processor scheduling tasks and the towers of Hanoi puzzle. From this we ascertain the size and functionality of the solutions provided. Prior to execution, the AEN has several key parameters to be initialised. Firstly the number of genes in the network. This is set between 15 and 20 to begin with and has a minimum value of 2 throughout optimisation. The networks are initialised with between 2 and 5 epigenetic switches with a minimum of 2 throughout optimisation.

The networks are optimised using NSGA II which has been shown to be very effective at evolving solutions within a multidimensional search space [3,4]. A total of 50 runs will be conducted for each experiment with a population size of 250 over 250 generations.

### 4.1   Coupled Inverted Pendulums

The coupled inverted pendulums task [5] consists of a set of 3 pendulums which are mounted to carts (1 per cart) on a 1-dimensional track. These carts are coupled together by a tether which restricts each cart's movement. The carts exist within a finite space, and must avoid the edges of this space. The objective of the task is to move the pendulums from the lower equilibrium position (swinging below the carts), and balance them in the upper unstable equilibrium position via the movements of the carts to which they are attached. To do this requires a periodic swinging movement to generate momentum in the pendulums; when the upper unstable equilibrium point is reached, the carts have to adapt their periodic behaviour to maintain it in that position.

The AEN controller is provided with 10 environmental inputs which refer to the position, momentum and angular velocity of a single cart and pendulum. These inputs are subject to a noise term to increase the difficulty of the task and to promote robustness in the evolved controllers. The AEN using this produces 2 outputs for each of the three carts. This is repeated 3 times for each time step. There is a total of 4000 time steps in total. There are 2 objective fitness values which represent the AEN's fitness. Firstly its objective performance, and secondly, an aggregate of its performance and the size of the network.

## 4.2 Processor Scheduling

The processor scheduling tasks simulates a multi core system and a set of tasks which need to be executed on it. There are 4 cores, each capable of executing a single task at a time. Each core is only capable of executing tasks for a set amount of time before its temperature exceeds a critical threshold of between 150 and 300 units (core 1 = 150, core 2 = 200, core 3 = 250, core 4 = 300). At each time step, if the core is currently executing its temperature will rise at 1 unit per time step. If the core is paused or inactive, the core will reduce its temperature by 1 unit per time step unless the temperature reaches 25, upon which it will remain static.

Each core can do one of 2 things at any given time. It can accept a task to process, and it can pause and un-pause any currently active core. If a core is active, it must finish processing its current task until completion, and cannot accept another task until that point. The AEN takes 4 inputs from each core and then produces 2 outputs. The four inputs are the core's temperature, its activity state and its pause state. The first of the 2 outputs controls whether a task is to be loaded on a core or the core's pause state is to be toggled. The second of the 2 outputs specifics which core the action is to be applied to.

There are 30 separate tasks to be scheduled. Each task is attributed a difficulty requiring either 50, 125 or 200 time steps to complete. A level of noise is introduced by taking these time step values from a Gaussian distribution. This is to provide noise to the simulation to improve the robustness of the AEN controllers. The task's difficulty is the number of time steps required to complete that task. The 3 different task difficulties are 50, 125 and 200. In total there are approximately 3000 consecutively executed time steps worth of tasks and the scheduler has 2000 time steps in which to have executed all tasks. The objective fitness values are the amount of tasks completed, the time steps required to do so, and an aggregate of the number of tasks completed and the network size. Only when all tasks are executed and run to completion will the scheduler stop before the 2000 steps are reached. At this point, the objective for the number of time steps automatically becomes active.

### 4.3   Towers of Hanoi

The towers of Hanoi is an NP hard mathematical puzzle [13]. The problem consists of 3 pegs and in this instance 5 discs of varying size, ordered from largest to smallest on the first peg. The objective is to move the discs from the first peg to the third peg. This is to be done without placing a larger disc upon a smaller one. The problem is represented as 2 sets of 3 variables. The first 3 describe the number of discs on each peg. The second 3 describe the sum of the weights of the dics on each peg, with 5 being the heaviest and 1 the smallest. The starting state is represented as [5,0,0,15,0,0] with the goal being [0,0,5,0,0,15]. The AEN will produce 2 outputs, the peg from which to move a disc, and the peg to move that disc to. If the move is invalid the task remains in its current state.

## 5   Results

For all tasks the AEN was able to find an optimal solution. In the coupled inverted pendulums, this was to maintain the pendulum in the upright position. For the scheduling task, this was to schedule all 30 tasks (of 3000 linear time steps in length) within 2000 time steps. For the Towers of Hanoi, all 5 discs were moved to the third peg.

For all tasks there was a distribution of results similar to that of the coupled inverted pendulums task shown in Fig. 4 where there is a general trade off between function and the size of the network. There is however an exception to this for the scheduling task where the highest performing network is also the smallest.

For all tasks an optimum solution is found containing under 6 genes or epigenetic switches. In every case this is below the sum of the number of inputs and outputs set by the task. It is to be noted that the number of genes and epigenetic switches within a network is typically lower than that shown in the results graphs. This is because there are often genes and epigenetic switches which exist within the network but are not connected either directly or via proxy to an output. Hence due to the small size of the AENs, they can easily be reduced in size further by very simple analysis.

### 5.1   Coupled Inverted Pendulums

There are a significant amount of instances of the AEN which are capable of producing the optimum behaviour. Of which, all of them contain either 9 units of less. This is 3 less than the sum of inputs and outputs set by the task. Of these instances there is a general trend of how the networks solve the problem. All of these instances use 2 specific sensors to control their behavior. These specify the position and angular velocity of the pendulum, and each of the networks which achieved the optimal behaviour used these inputs. In addition, all these networks utilise at least one epigenetic switch.

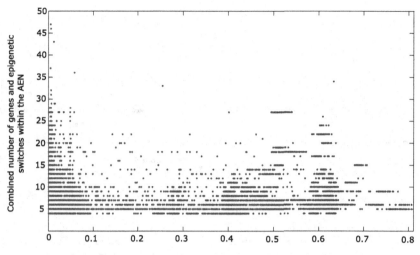

Fig. 4. The size of the networks ($y$ axis - lower numbers are better) plotted against the performance of the networks ($x$ axis - higher numbers are better) for all AEN's at the final generation. The performance is calculated as the normalised amount of time steps the pendulums were positioned in the upright position. The threshold for the AEN developing an optimal balancing behavior is above 0.7.

When the size of the AENs reaches a threshold, it becomes possible to mathematically reproduce their behavior in a set of equations. An example of the smallest network which can optimally solve the task is shown in Eq. 3 (It is to be noted that this network contained 3 genes and 2 epigenetic switches However, one of each was unconnected, and has been removed). This equation fully describes each functional gene within the network to 2 decimal places where $S_0$ and $S_9$ are sensor inputs from the task specifying the speed and location of the pendulum.

$$Gene\ 1 = \frac{1}{1 + e^{(-7.90 \cdot S_0 \cdot (0.63)) - 0.76}} \qquad Gene\ 2 = \frac{1}{1 + e^{(-6.38 \cdot S_9 \cdot (-0.48)) - 0.74}}$$

$$Gene\ 3 = \begin{cases} 0.45 & \text{if} \quad \dfrac{1}{1 + e^{(-12.90 \cdot (gene3)) - 0.07}} < 0.5 \\ 0 & \text{otherwise} \end{cases}$$

$$(3)$$

## 5.2 Processor Scheduling

Of network structures capable of solving this task, the majority showed very little correlation between structure and function. Some instances of the AEN

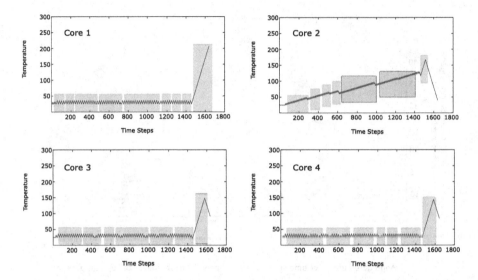

**Fig. 5.** The temperatures of all 4 cores with all 30 tasks successfully scheduled in 1653 steps. The grey boxes denote the individual tasks and how long they were executed for in each core. In cores 2, 3 and 4 it can be seen that for the majority of execution, the temperature is significantly lower then the threshold for damage (core 1 = 150, core 2 = 200, core 3 = 250, core 4 = 300).

solve the task utilising the majority of the environmental inputs. Approximately 40 % use epigenetic switching to achieve this. However, there are also instances which function well which utilise no inputs from the environment and use no epigenetic switches.

In general, the overall functionality in which the AEN uses to solve the task is to create an oscillator which allows execution of a task for a small amount of time, and then pauses the executing core for a set amount of time. This allows for reasonably efficient execution of the tasks without significantly increasing the temperatures of the cores.

Equation 4 shows a very efficient instance of the AEN which is able to schedule all 30 tasks. There was a single gene and 2 epigenetic switches within this network which were removed as they were not connected to any other units. This is the mathematical representation of the the AEN shown to be scheduling the tasks in Fig. 5.

$$Gene \ 1 = \frac{1}{1 + e^{(-8.04 \cdot (gene2 + gene3)) - 0.14}} \qquad Gene \ 2 = \frac{1}{1 + e^{(-7.39 \cdot (gene1 + gene3)) - 0.31}}$$

$$Gene \ 3 = \frac{1}{1 + e^{(-6.82 \cdot (gene1 + gene2)) - 0.30}}$$

$$(4)$$

## 5.3    Towers of Hanoi

The towers of Hanoi task is different from that of the previous two in that it is known to have a recursive solution. This is demonstrated by the analysis of the AEN's which are able to solve it, with almost all instances having evolved solutions which contain no inputs.

$$Gene\ 1 = \frac{1}{1 + e^{(-19.58\cdot(gene2+gene3))}-0.08} \qquad Gene\ 2 = \frac{1}{1 + e^{(-11.8\cdot(gene1+gene3))}-0.36}$$

$$Gene\ 3 = \frac{1}{1 + e^{(-13.3\cdot(gene1+gene2))}-0.08}$$

$$(5)$$

Although many of the networks which solve the task use 6 or more genes or epigenetic molecules, many of these units are surplus to the functionality of the network, and a significant number of solutions only contain 3 functional units. A mathematical example of which can be seen in Eq. 5. The best solutions are able to solve the problem in about 150 moves (where 31 is optimal), however the AEN's also try a significant number of invalid moves which accounts for usually the same number of valid moves. Hence, for 150 valid moves, the AENs will usually try around 300 moves.

## 6    Conclusions

In this work we have shown that the adaptations to the AEN give it the capabilities of autonomously evolving efficient solutions to complex tasks. The AENs produced solutions to all of the tasks outlined which contain a combination of 6 or less genes and epigenetic switches. With very simple post processing, the most efficient solutions contained a combination of only 3 epigenetic switches, whilst maintaining functionality. In addition, this work also adds weight to the idea that the AEN is a general processing tool, capable of solving a wide range of problems.

In future work, we plan to investigate how the solutions produced by the AEN can be encapsulated within electronic circuits. Additionally, we plan to investigate how apparent biological properties such as genetic redundancy are within the AEN structures.

**Acknowledgements.** The authors would like to thank the EPSRC for their support of this work through the Platform Grant EP/K040820/1. Data created during this research is available at the following DOI: 10.15124/3f245e80-c306-4ada-8920-a0282e4962b3.

# References

1. Bird, A.: Perceptions of epigenetics. Nature **447**(7143), 396–398 (2007)
2. Bull, L.: Consideration of mobile DNA: new forms of artificial genetic regulatory networks. Nat. Comput. **12**(4), 443–452 (2013)
3. Deb, K., Agrawal, S., Pratap, A., Meyarivan, T.: A fast elitist non-dominated sorting genetic algorithm for multi-objective optimization: NSGA-II. In: Deb, K., Rudolph, G., Lutton, E., Merelo, J.J., Schoenauer, M., Schwefel, H.-P., Yao, X. (eds.) PPSN 2000. LNCS, vol. 1917, pp. 849–858. Springer, Heidelberg (2000)
4. Deb, K., Pratap, A., Agarwal, S., Meyarivan, T.: A fast and elitist multiobjective genetic algorithm: NSGA-II. IEEE Trans. Evol. Comput. **6**(2), 182–197 (2002)
5. Hamann, H., Schmickl, T., Crailsheim, K.: Coupled inverted pendulums: a benchmark for evolving decentral controllers in modular robotics. In: Proceedings of the 13th Annual Conference on Genetic and Evolutionary Computation, pp. 195–202. ACM (2011)
6. Hinton, G., Deng, L., Yu, D., Dahl, G.E., Mohamed, A., Jaitly, N., Senior, A., Vanhoucke, V., Nguyen, P., Sainath, T.N., et al.: Deep neural networks for acoustic modeling in speech recognition: the shared views of four research groups. IEEE Sig. Process. Mag. **29**(6), 82–97 (2012)
7. Huang, S.: Reprogramming cell fates: reconciling rarity with robustness. Bioessays **31**(5), 546–560 (2009)
8. Lones, M.A., Turner, A.P., Fuente, L.A., Stepney, S., Caves, L.S.D., Tyrrell, A.M.: Biochemical connectionism. Nat. Comput. **12**(4), 453–472 (2013)
9. Lones, M.A., Tyrrell, A.M., Stepney, S., Caves, L.S.: Controlling complex dynamics with artificial biochemical networks. In: Esparcia-Alcázar, A.I., Ekárt, A., Silva, S., Dignum, S., Uyar, A.Ş. (eds.) EuroGP 2010. LNCS, vol. 6021, pp. 159–170. Springer, Heidelberg (2010)
10. Masel, J., Trotter, M.V.: Robustness and evolvability. Trends Genet. **26**(9), 406–414 (2010)
11. Moran, L.A., Horton, H.R., Scrimgeour, G., Perry, M.: Principles of Biochemistry. Pearson, Boston (2012)
12. Oquab, M., Bottou, L., Laptev, I., Sivic, J.: Learning and transferring mid-level image representations using convolutional neural networks. In: 2014 IEEE Conference on Computer Vision and Pattern Recognition (CVPR), pp. 1717–1724. IEEE (2014)
13. Reid, C.R., Sumpter, D.J., Beekman, M.: Optimisation in a natural system: Argentine ants solve the Towers of Hanoi. J. Exp. Biol. **214**(1), 50–58 (2011)
14. Reil, T.: Dynamics of gene expression in an artificial genome - implications for biological and artificial ontogeny. In: Floreano, D., Mondada, F. (eds.) ECAL 1999. LNCS, vol. 1674, pp. 457–466. Springer, Heidelberg (1999)
15. Turner, A.P.: The Artificial Epigenetic Network. Ph.D. thesis, University of York (2013)
16. Turner, A.P., Lones, M.A., Fuente, L.A., Stepney, S., Caves, L.S.D., Tyrrell, A.M.: The artificial epigenetic network. In: 10th Internation Conference on Evolvable Systems, Singapore, pp. 66–72. IEEE Press, April 2013
17. Turner, A.P., Lones, M.A., Fuente, L.A., Tyrrell, A.M., Stepney, S., Caves, L.S.D.: Controlling complex tasks using artificial epigenetic regulatory networks. BioSystems **112**(2), 56–62 (2013)
18. Wang, R.S., Saadatpour, A., Albert, R.: Boolean modeling in systems biology: an overview of methodology and applications. Phys. Biol. **9**(5), 055001 (2012)

19. Yang, X.S., Cui, Z., Xiao, R., Gandomi, A.H., Karamanoglu, M.: Swarm intelligence and bio-inspired computation: theory and applications. Newnes (2013)
20. Zhou, A., Qu, B.Y., Li, H., Zhao, S.Z., Suganthan, P.N., Zhang, Q.: Multiobjective evolutionary algorithms: a survey of the state of the art. Swarm Evol. Comput. $1(1)$, 32–49 (2011)

# Metabolomics and Phenotypes

# Sensitivity of Contending Cellular Objectives in the Central Carbon Metabolism of *Escherichia Coli*

Max Sajitz-Hermstein[(✉)] and Zoran Nikoloski

Systems Biology and Mathematical Modeling Group, Max Planck Institute
of Molecular Plant Physiology, Am Mühlenberg 1, 14476, Potsdam, Germany
sajitz@mpimp-golm.mpg.de

**Abstract.** To ensure homeostasis as well as proliferation, cellular systems usually adapt to changes in environmental and intracellular conditions at the level of the flux phenotype. The latter is characterized by the biochemical reaction rates in the underlying metabolic network and depends on the concentration of individual metabolites. As a result, concentrations of metabolites with large effect on the flux phenotype are expected to be tightly controlled. We examine the sensitivity of the flux phenotype upon changes in metabolite concentrations via the shadow prices in a flux balance analysis using multiple contending objectives of the central carbon metabolism of E. coli. The shadow prices of the metabolites are determined individually for sampled solutions of the Pareto front and objective functions. Utilization of 13C flux measurements for different environmental conditions enables us to draw conclusions about the relation of shadow prices and physiological cellular states. We find that E. coli operates in the vicinity of an area of the Pareto front which exhibits low variation of shadow prices compared to the whole front, which enables to react to changing conditions without large changes in the reguatory machinery. In addition, the determined shadow prices under different conditions suggest an increased requirement for regulation of concentrations of metabolites from the pentose phosphate pathway under carbon-limiting conditions compared to aerobe conditions. Our study extends the applicability of concepts from classical constraint-based modelling in a multi-objective settings to obtain predictions about regulation of metabolite levels based solely on stoichiometry.

**Keywords:** Flux balance analysis · Shadow prices · Multi-objective optimization · Sensitivity · Escherichia coli · Central carbon metabolism

## 1 Introduction

Biological systems perpetually sense and respond to changes in environmental and intracellular conditions. To ensure homeostasis as well as proliferation, cellular systems usually adapt to the experienced changes via an integrated response of metabolic and regulatory networks manifested at the level of the flux phenotype. The latter is characterized by the biochemical reaction rates in the

© Springer International Publishing Switzerland 2015
M. Lones et al. (Eds.): IPCAT 2015, LNCS 9303, pp. 169–172, 2015.
DOI: 10.1007/978-3-319-23108-2_14

underlying metabolic network and depends on the concentration of individual metabolites. As a result, concentrations of metabolites with large effect on the flux phenotype are expected to be tightly controlled.

Shadow prices describe the sensitivity of the objective function of a linear program upon perturbation of individual constraints. In a cellular setting, shadow prices can be used to analyze the change of the optimal flux phenotype upon introduction/removal of quantities of a given metabolite in the framework of flux balance analysis [3,4,7]. It has been shown that shadow prices are suitable predictors of temporal variation of metabolite concentrations as well as indicators of the growth-limiting effects of the respective metabolites [2]. Therefore, shadow prices can be used to examine the requirement for regulating individual metabolite concentrations solely based on a stochiometric model of the metabolic network.

To the best of our knowledge, existing studies have concentrated only on shadow prices for a single objective function. However, it was shown that utilization of a combination of multiple objective functions can improve the accuracy of the predicted flux phenotypes [5]. For the case of *Escherichia coli,* a multi-objective analysis established that flux phenotypes determined under different environmental conditions could be best described by individual weightings of objective functions [6]. Moreover, it was found that flux phenotypes are in very close vicinity to the Pareto front describing the set of noninferior solutions, *i.e.,* flux phenotypes which can only improve one objective at the price of reducing at least one other objective.

Here, we examine the shadow prices in an analysis using multiple contending objectives of the central carbon metabolism of *E. coli* in accordance with the work of [6]. The shadow prices of the metabolites are determined individually for sampled solutions of the Pareto front and objective functions. Utilization of $^{13}$C flux measurements for aerobe as well as carbon- and nitrogen-limited conditions allows us to determine the section of the Pareto front in which the cell operates under the examined conditions. Therefore, the combination of experimentally estimated fluxes and multi-objective analysis enabled us to draw conclusions about the relation of shadow prices and physiological cellular states.

Our results indicate that *E. coli* operates in the vicinity of an area of the Pareto front which exhibits low variation of shadow prices compared to the whole front. This finding implies that *E. coli* is able to react upon changes in environmental conditions without large changes in the regulatory machinery. In addition, the determined shadow prices under different conditions suggest an increased requirement for regulation of concentrations of metabolites from the pentose phosphate pathway under carbon-limiting conditions compared to aerobe conditions.

## 2    Materials and Methods

The Pareto front describes the set of noninferior solutions of a multi-objective programming problem. A representative set of solutions can be determined by successive sampling of solutions which cover the entire Pareto front. We determine a representative set of the Pareto front solutions of *E. coli*'s central carbon metabolism in accordance with [6], utilizing the objective functions of maximizing biomass yield, maximizing ATP yield, and minimizing total flux. The representative subset of the Pareto front is determined via the epsilon-constraint method [1]. Shadow prices are then individually calculated for each element of the representative set and each objective function. To this end, we fix all but one of the objectives to the values of an element of the representative set, and calculate the effects of perturbing the steady-state constraints (corresponding to the individual metabolites) on the remaining objective. Elements of the representative subset that are closest to the experimentally estimated flux phenotypes are determined by optimization as described by [6].

## 3   Results

We compare shadow prices across the entire Pareto front as well as physiological flux phenotypes. Moreover, we provide an analysis of shadow prices for physiological flux phenotypes to obtain insights into the adaption of regulatory mechanisms employed upon environmental changes.

Opposing changes of shadow prices with respect to individual objectives upon shifts in environmental conditions indicate that concurrent regulatory mechanisms may have to be utilized. Based on Kendall's rank correlation coefficient, we find that shadow prices across the experimentally determined flux phenotypes are highly correlated for each two of the three objectives and for most of the metabolites. In contrast, correlations are lower across the whole Pareto front and less metabolites show high correlations. We conclude that *E. coli* operates near an area of the Pareto front enabling adaptation to environmental changes with low requirement of concurrent mechanisms for regulating the concentration of individual metabolites.

The comparison of shadow prices obtained for different environmental conditions identifies increasing and decreasing sensitivity to changes in metabolite concentrations. In particular, we find that metabolites of the pentose phosphate pathway exhibit larger sensitivity under conditions of carbon- and nitrogen-limitation in comparison to aerobe conditions. This finding indicates that metabolites of the pentose phosphate pathway should be more regulated under stress conditions. Altogether, our study extends the applicability of concepts from classical constraint-based modeling in a multi-objective settings to obtain predictions about regulation of metabolite levels based solely on stoichiometry.

## References

1. Miettinen, K.: Nonlinear Multiobjective Optimization. Kluwer Academic Publishers, Boston (1999)
2. Reznik, E., Mehta, P., Segrè, D.: Flux imbalance analysis and the sensitivity of cellular growth to changes in metabolite pools. PLoS Comput. Biol. **9**(8), e1003195 (2013)

3. Savinell, J., Palsson, B.: Network analysis of intermediary metabolism using linear optimization. I. Development of mathematical formalism. J. Theor. Biol. **154**, 421–454 (1992)
4. Savinell, J.M., Palsson, B.O., Arbor, A.: Network analysis of intermediary metabolism using linear optimization. II. Interpretation of hybridoma cell metabolism the uses of linear optimization theory to calculate and interpret fluxes in metabolic. J. Theor. Biol. **154**, 455–473 (1992)
5. Schuetz, R., Kuepfer, L., Sauer, U.: Systematic evaluation of objective functions for predicting intracellular fluxes in *Escherichia coli*. Mol. Syst. Biol. **3**, 119 (2007)
6. Schuetz, R., Zamboni, N., Zampieri, M., Heinemann, M., Sauer, U.: Multidimensional optimality of microbial metabolism. Science **336**, 601–603 (2012)
7. Varma, A., Palsson, B.O.: Metabolic flux balancing: basic concepts scientific and practical use. Nat. Biotechnol. **12**(10), 994–998 (1994)

# Towards a Graph-Theoretic Approach to Hybrid Performance Prediction from Large-Scale Phenotypic Data

Alberto Castellini[1,2]([⊠]), Christian Edlich-Muth[1,2], Moses Muraya[3],
Christian Klukas[3], Thomas Altmann[3], and Joachim Selbig[1,2]

[1] Bioinformatics Group, Institute for Biochemistry and Biology,
University of Potsdam, 14476 Potsdam, Germany
{alberto.castellini,christian.edlich-muth}@uni-potsdam.de
[2] Max Planck Institute of Molecular Plant Physiology, 14476 Potsdam, Germany
selbig@mpimp-golm.mpg.de
[3] Department of Molecular Genetics, Leibniz Institute of Plant Genetics and Crop
Plant Research (IPK) Gatersleben, Stadt Seeland, Germany
{muraya,klukas,altmann}@ipk-gatersleben.de

**Abstract.** High-throughput biological data analysis has received a large
amount of interest in the last decade due to pioneering technologies that
are able to automatically generate large-scale datasets by performing millions of analytical tests on a daily basis. Here we present a new network-
based approach to analyze a high-throughput phenomic dataset that was
collected on maize inbreds and hybrids by an automated phenotyping
facility. Our dataset consists of 1600 biological samples from 600 different genotypes (200 inbred and 400 hybrid lines). On each sample, 141
phenotypic traits were observed for 33 days. We apply a graph-theoretic
approach to address two important problems: *(i)* to discover meaningful patterns in the dataset and *(ii)* to predict hybrid performance in
terms of biomass based on automatically collected phenotypic traits.
We propose a modelling framework in which the prediction problem
becomes transformed into finding the shortest path in a correlation-based
network. Preliminary results show small but encouraging correlations
between predicted and observed biomass. Extensions of the algorithm
and applications of the modelling framework to other types of biological
data are discussed.

## 1 Introduction

In the continuous effort to obtain improved crop plant varieties, high-throughput
methods have taken center stage in recent years. The methods to assess genomes
and metabolomes, to take just two examples, are fully established and standard-
ized protocols and processing software are available. This is not the case for
phenomic data, which is the large and parallel collection of phenotypic information on a large number of samples.

---

Alberto Castellini and Christian Edlich-Muth contributed equally to this work.

© Springer International Publishing Switzerland 2015
M. Lones et al. (Eds.): IPCAT 2015, LNCS 9303, pp. 173–184, 2015.
DOI: 10.1007/978-3-319-23108-2_15

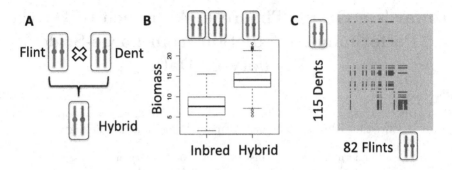

**Fig. 1.** The hybrid principle. A: A Dent and a Flint inbred line (green and red respectively) combine in a cross to give rise to a hybrid. Both Dent and Flint are taken to be homozygous for each locus, therefore their two chromosomes are identical. B: hybrid vigour (heterosis). Hybrids are vastly superior in performance measures such as biomass. C: hybrid space. The panel of nearly 200 inbred lines could give rise to about 10,000 hybrids. The blue dots on the grey background represent the 400 hybrids that were experimentally tested. Thus only 4 % of hybrid space is probed (Color figure online).

Phenomic data usually implies taking a large number of pictures of a plant from different angles and at different wavelengths during an extended period of development. The images are automatically processed and features, e.g. plant height and volume, are extracted, yielding a three-dimensional matrix of samples, features and time [3,13]. In order to restrict human interference to a minimum and to make sure each plant is watered and phenotyped regularly, fully automated robotic greenhouses have been constructed in which the plants are moved on conveyor belts around the greenhouse until it is their turn to be weighed, watered and imaged [8].

The particular biological question that motivates our work is how phenomics of a large variety of maize cultivars could be used in aiding the selection of superior varieties for agricultural use. Maize is the worlds third most important crop plant (after wheat and rice) and has an impressive track record in improved yields for more than eight decades. The technology behind this is to cross two homozygous varieties (inbreds) yielding a highly homogenous population of heterozygous offspring, so-called hybrids, which are far superior to their parents. This effect is widely known as heterosis [1,6,15] and is illustrated in Fig. 1.

Since the number of possible hybrids that can be generated from panels of inbred lines (hybrid space) is the product of the number of parental male and female lines (tens to hundreds of thousands potential crosses each year) it is not possible to test each of them even in the most rudimentary fashion which has motivated the development of computational methods that predict the desired properties. While prediction is of course likely to be very error-prone, it can at the very least help to considerable reduce in size the part of hybrid space that needs to be experimentally evaluated.

Several statistical and machine learning methods have been proposed to this end, some of them focusing on genomic prediction [14,19], others integrating both genetic and metabolic profiles [5,16]. The majority of these approaches make use of both feature selection and regression techniques for reducing the number of selected biomarkers and generating more powerful and interpretable models. Classical approaches, such as partial least squares [18], and more recent ones, like regularization and LASSO-based techniques [7], have proved to reach good performance in this field. Alternative approaches [4] also exist where, for instance, biomass/heterosis prediction is treated as a classification problem and methods such as support vector machines, linear discriminant analysis and random forests are used to generate predictive models [7].

In this paper, we take a graph-theoretical approach [17]. After presenting an exploratory overview of the data we convert them into a correlation network and analyze the properties thereof. Finally, we develop an algorithm that explores the correlation network to predict properties (performance) of candidate hybrids from the same breeding program. Section 2 formally describes the main features of our dataset. The methodology (i.e., data preparation, exploratory analysis and predictive breeding approach) is presented in Sect. 3 and preliminary results are discussed in Sect. 4. In Sect. 5 we propose possible extensions to this approach.

## 2   Dataset

In our study, 197 inbred lines derived from European and North American varieties form the two so-called heterotic pools, "Dent" and "Flint" (sets $\mathcal{D}$ and $\mathcal{F}$ in Definition 1, below). The hybrids in this study are the F1 generation of a directional cross between a Dent and a Flint inbred line, such that the male parent is always a Flint. Two subsets of respectively 26 Dents and 22 Flints were selected as founders (sets $\mathcal{D}'$ and $\mathcal{F}'$ in Definition 1) of the cross and 392 Dent x Flint hybrids were made. The number of hybrids is a mere 4 % of the hybrid space spanned by the 115 Dents and 82 Flints, but 58 % of all possible Dent x Flint hybrids derived from the founders. Hereafter, we shall refer to Dents, Flints and hybrids as the three genetic pools or simply "pools".

Plants were grown from seeds and were transferred to the robotic platforms as seedlings. Four seedlings of the same genotype were place in a carrier. Each plant was imaged by a visible-range, UV and infrared camera from several angles. From the raw image files, features of interest were derived as described in [9,10]. From over 300 features, the most reproducible 141 were selected. They describe either spectral (e.g. sum of side view fluorescence in the chlorophyll range) or structural (e.g. visible spectrum side view hull area) properties. Imaging started 13 or 15 days after sowing (DAS) and the initial pictures are an average over all four plants. Two of the plants were harvested 28 DAS to manually determine the biomass. The remaining two plants were imaged together and harvested at the end of the experiment, 48 days after sowing. In each greenhouse run up to 400 carriers (genotypes) can be processed. The data of this study comprise four greenhouse runs: two with inbreds and two with hybrids.

The four greenhouse runs were combined into a single dataset, a matrix of 67425 rows and 141 columns. Each row represents the observations on a specific sample (i.e., carrier) of one of the four greenhouse runs on a specific day after sowing (DAS). A total of 1584 plants were observed for 34 to 36 days. For each of the 197 inbred genotypes there were four independent biological samples, and for each of the 392 hybrids there were two independent samples. In addition, final biomass measurements (wet weight) on all genotypes was available (modelled as function $\gamma(\cdot)$ in Definition 1 below).

**Data Preprocessing and Cubes Generation.** After median normalisation by greenhouse run, the values of phenotypic traits of plants of the same genotype were averaged yielding a matrix of 20404 rows (i.e., one row for each genotype-DAS pair) and 141 columns (i.e. one column for each phenotypic trait). Since inbred DAS range from 13 to 48 and hybrid DAS from 15 to 48, we focused our analysis on the common DASs only, namely 15–48. The analysis of missing values showed that 66 % of hybrid phenotypes of DAS 15 and 13 % of hybrid phenotypes of DAS 18 were missing. We therefore also removed DAS 15 from our analysis and imputed missing values of DAS 18 with random forests [2,11]. The final dataset is summarized in Fig. 2 as three genotype-feature-time cubes, one for Dents (115 genotypes of which 26 are founders), one for Flints (82 of which 22 are founders) and one for hybrids (392 genotypes). All three cubes have 141 phenotypic traits and 33 DAS (from 16 to 48). The subdivision into three sub-cubes is motivated by the genetic relationship that leads to the formation of a hybrid from exactly one Dent and one Flint. This relationship, a two-generation pedigree, can be interpreted as a function $\delta$ that maps (Dent, Flint) pairs to hybrids. The pedigree is rendered as a two-dimensional map in Fig. 2.

In the following we introduce some mathematical notation, then used throughout the manuscript, to formally describe the main features of our dataset.

**Definition 1 (Dataset).** The dataset under investigation (see Fig. 2) includes three 3-dimensional arrays of observed phenotypic traits

$$M_{\mathcal{D}} : \mathcal{D} \times \mathcal{P} \times \mathcal{T} \to \mathbb{R}^+, \quad M_{\mathcal{F}} : \mathcal{F} \times \mathcal{P} \times \mathcal{T} \to \mathbb{R}^+, \quad M_{\mathcal{H}} : \mathcal{H} \times \mathcal{P} \times \mathcal{T} \to \mathbb{R}^+,$$

a *breeding function* which maps couples of genotypes to hybrid genotypes

$$\delta : \mathcal{D}' \times \mathcal{F}' \to H$$

and a *genotype performance* mapping which associates a performance measure (e.g., biomass value) to each genotype

$$\gamma : \mathcal{D} \cup \mathcal{F} \cup \mathcal{H} \to \mathbb{R}^+,$$

where:

- $\mathcal{P} = \{p_1, \ldots, p_n\}$ is a set of *phenotypic traits*,

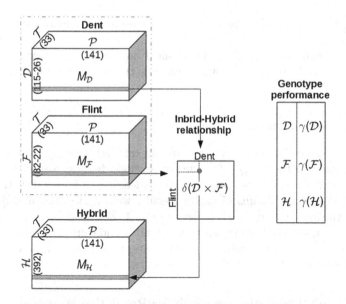

**Fig. 2.** Preprocessed dataset. The data are subdivided into three cubes according to the genetic pools (Dent, Flint and hybrid). The Dent pool contains 115 lines of which 26 are founders, the Flint pool contains 82 lines of which 22 are founders and the hybrid pool contains 392 lines. The number of points on the time axis (33) and the number of features (141) is the same in each cube. One Dent and one Flint are the parents of one hybrid. This relationship is captured in the pedigree which is represented here as a two-dimensional mapping from parental inbreds to hybrid.

- $\mathcal{D} = \{d_1, \ldots, d_m\}$ is a set of *Dent genotypes*,
- $\mathcal{D}' \subseteq \mathcal{D} \mid \mathcal{D}' = \{d_1, \ldots, d_{m'}\}, m' \leq m$ is a subset of Dent genotypes tested for hybridization,
- $\mathcal{F} = \{f_1, \ldots, f_r\}$ is a set of *Flint genotypes*,
- $\mathcal{F}' \subseteq \mathcal{F} \mid \mathcal{F}' = \{f_1, \ldots, f_{r'}\}, r' \leq r$ is a subset of Flint genotypes tested for hybridization,
- $\mathcal{H} = \{h_1, \ldots, h_s\}$ is a set of *hybrid genotypes*,
- $\mathcal{D} \cap \mathcal{F} \cap \mathcal{H} = \emptyset$,
- $\mathcal{T} = \{t_1, \ldots, t_q\}$ is a set of natural numbers representing *days after sowing (DAS)*.

The correlation (based on phenotypic traits) between couples of genotypes observed at the same time can be computed, according to the the notation defined above, by the function $\rho : (\mathcal{D} \cup \mathcal{F} \cup \mathcal{H}) \times (\mathcal{D} \cup \mathcal{F} \cup \mathcal{H}) \times \mathcal{T} \to [-1, 1] \in \mathbb{R}$.

## 3   Method

### 3.1   Problem Definition

In this work we address two important problems:

1. to discover meaningful patterns in the correlation structure of cubes $M_{\mathcal{D}}$, $M_{\mathcal{F}}$ and $M_{\mathcal{H}}$,
2. to rank all couples of available Dent and Flint genotypes according to some measure (based on biomass) which enables the prediction of hybrid performance.

## 3.2 Data Normalization

The range of absolute values of each column differed strongly from feature to feature. Moreover, the range of values of a single feature was in some cases very different in hybrids compared to inbreds leading to bimodal distributions. For this reason, the columns of each sub-cube were scaled and mean-centered separately (mean of 0 and standard deviation of 1, Z-scores). This normalization is necessary to make the features more comparable and thus usable for generating networks of genotypes. The effect of normalization on the correlation histograms is illustrated in Fig. 3.

## 3.3 A Graph-Theoretic Approach for Predictive Breeding

Given the three cubes of Fig. 2 three kinds of correlation structures can be investigated, namely with respect to genotypes, phenotypes and time. In the first case the aim is to understand the relationships among genotypes, based on phenotypic traits. We observe that the mapping $\delta$ between pairs of inbreds (i.e., one Dent

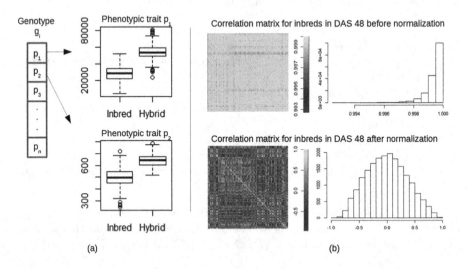

**Fig. 3.** Data normalization. (a) Two examples of heterosis in phenotypic traits. The range of values is very different for inbreds and hybrids and would lead to a bimodal distribution if the pools were combined. (b) Heat map and histograms of the correlation matrix for all inbred genotypes (i.e., dent and flint) on DAS 48 with or without (bottom and top panels respectively) normalization.

and one Flint) and hybrids can enrich the analysis based on genotypes because it connects patterns in the correlation structure of inbreds with patterns in the correlation structure of hybrids. In the following we will therefore focus on the analysis of genotypes and only briefly touch on correlations in time and between phenotypes.

There exist several methods of network inference, some of them developed in the context of gene regulatory networks, where high-throughput data are also available [12]. Here we use a correlation-based network, which allows us to reformulate the problem of predicting hybrid performance as the search for the shortest path between two nodes.

**Definition 2 (Network $\mathcal{N}_t$).** Network $\mathcal{N}_t$ is a graph $(V, E)$ where:

- $V = \mathcal{D} \cup \mathcal{F} \cup \mathcal{H}$ is the set of nodes,
- $E = \{e_{ij} \mid i, j \in V\}$ is the set of edges having weights $w_{ij}$ defined as follows:
  - $w_{ij} = 1 - \rho(i, j, t)$, if $(i \in \mathcal{D} \ \wedge \ j \in \mathcal{D}) \ \vee \ (i \in \mathcal{F} \ \wedge \ j \in \mathcal{F}) \ \vee \ (i \in \mathcal{H} \ \wedge \ j \in \mathcal{H})$, where $\rho(i, j, t)$ is the correlation between genotype $i$ and genotype $j$ at day $t$ after sowing.
  - $w_{ij} = 10$, if $i \in \mathcal{H}$ and $j \in \delta^{-1}(i)$, where $\delta$ is the breeding function,

Figure 4 outlines an example of network $\mathcal{N}_t$ which combines the intra-pool correlation structure with the breeding function. Before adding the edges of the breeding function (those with weight equals to 10), all nodes in each connected component are from the same pool and the network consists of three connected components, one for Dents, one for Flints and one for hybrids.

The distance between two genotypes (i.e., nodes) of the same pool in a specific day $t$ is defined as 1 minus the correlation coefficient between them, given their phenotypic values in $t$, and is encoded by edge weights. Edges originating from the breeding function have a weight equals to 10, which is higher than the maximum intra-pool weight. This value avoids shortest paths going back and forth from Dents/Flints to hybrids.

The usage of breeding function edges and distance edge weights has a number of consequences: (i) a path $p_{df}$ between a given Dent $d$ and a given Flint $f$ must exist if at least one hybrid exists; (ii) the length of the path (which is related to prediction quality) is the sum of its edge weights, which are higher the less correlated the connected nodes are; (iii) every path $p_{df}$ must contain at least one hybrid.

The nodes of the network $\mathcal{N}_t$ can be tagged with additional information. In the following we describe how this property could be exploited to predict how much biomass a fully-grown plant yields at harvest. The task of hybrid prediction for our purposes is defined in the following way: to accurately predict the performance of candidate hybrids given that *(i)* the parents of the hybrid are nodes in $\mathcal{N}_t$, *(ii)* some measurement of hybrid performance on all hybrid nodes in $\mathcal{N}_t$ is available.

Let us therefore assume that we have collected a performance measure (e.g., final biomass) on all hybrids in our breeding program. We want to predict the performance of a candidate hybrid $h = \delta(d, f)$ which could be realized by crossing

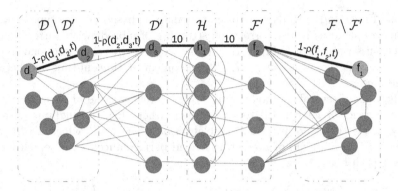

**Fig. 4.** Network $\mathcal{N}_t$. The graph is schematically subdivided into Dent, Hybrid and Flint nodes. An edge within these components represents the distance, in terms of correlation, between the nodes (0: maximum positive correlation, 1: no correlation, 2: maximum negative correlation). An edge between an inbred and a hybrid results from the breeding function. The illustrated path from $d_1$ reaches $f_1$ via a Dent in the founder layer $\mathcal{D}'$, a hybrid ($h_1 = \delta(d_3, f_2)$) and a Flint in the founder layer $\mathcal{F}'$. In our predictive framework, this path (taken to be the shortest path) provides a means to estimate the properties of the potential hybrid "$d_1 \times f_1$" which is not a node in the graph (otherwise the appropriate edge would have been present in the graph).

the Dent $d$ with the Flint $f$. Both $d$ and $f$ are nodes in $\mathcal{N}_t$ while $h$ is not a node in $\mathcal{N}_t$. The idea how to exploit the network (correlation) structure is this: if a path from $d$ to $f$ exists in the network, then the hybrids in that path for which the performance measure is known can be used to predict the performance of $h = \delta(d, f)$. The idea is based on the fact that any edge in the network guarantees a degree of similarity (correlation) between the two nodes based on all phenomic features. Therefore it is reasonable to expect a similar amount of correlation also for the performance measure which is a phenotypic trait as well. By using highly correlated genotypes as stepping-stones and the breeding function as a bridge between the pools, we can get from $d$ to $f$. Should several such paths exist, the shortest path is to be used for prediction. Moreover, if more than one shortest path exists, all hybrids in these paths are averaged. The algorithm in Table 1 sketches the main steps of the process of prediction and quality assessment of hybrid performance using this approach.

**Table 1.** Predictive breeding algorithm

| Predictive breeding $(\mathcal{D}, \mathcal{F}, \mathcal{N}_t)$ |
| --- |
| 1. For each couple $(d, f)$ $\mid$ $d \in \mathcal{D}, f \in \mathcal{F}$ |
| 2. Compute the shortest path between $d$ and $f$ in the network $\mathcal{N}_t$ |
| 3. *Performance measure* $\varphi(d, f)$: average value of biomasses of hybrids in this path |
| 4. *Quality measure* $\phi(d, f)$: path length (the smaller the better) |

## 4   Results

### 4.1   Biomass Performance Prediction

The predictive breeding algorithm was run to calculate the performance measure $\varphi(d, f)$ (biomass) on all $\mathcal{D} \times \mathcal{F}$ pairs in $\mathcal{N}_{48}$, the final time point. It is important to note that the predicted value of the biomass of one of the known hybrids is always equal to its observed one because the shortest path from $d'$ to $f'$ is directly via the hybrid $\delta(d', f')$. The median path length between unknown $(d, f)$ pairs is 0.6 and the median number of nodes in each path is 5. Thus, a representative path starts at a non-founder $d$, continues via a Dent founder $d'$, a hybrid $h = \beta(d', f')$, a Flint founder $f'$, and ends in $f$. The number of hybrids in the path is almost always 1 - therefore our performance measure is simply the biomass of that hybrid. In the few cases in which two hybrids were found in the shortest path biomass is the average of both.

In order to validate the algorithm, we deleted the edges of each hybrid $h = \delta(d', f')$ in turn and calculated the performance measure $\varphi(d', f')$ in this network. This is essentially leave-one-out cross-validation. The predicted versus observed hybrid biomasses are shown in Fig. 5. The correlation between observed and expected biomass is 0.35. Given that we are using a very simple algorithm that can be improved in numerous ways, this is an encouraging result. Interestingly, the path length (the quality measure $\phi(d, f)$) versus the log-ratio of predicted and observed biomass shows no indication that a longer path length is worse at predicting hybrid performance. On the other hand, in absolute terms the

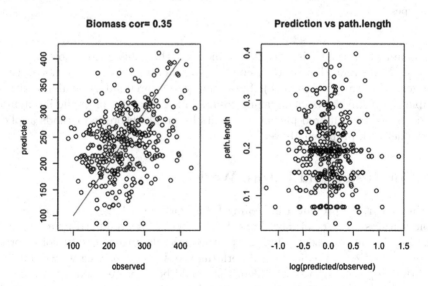

**Fig. 5.** Validation of the algorithm. Performance of known hybrids $h \in \mathcal{H}$ were predicted after removing all of their edges in turn. The scatter plot predicted vs observed is shown in the left panel. In the right panel, the prediction result is compared to the path length. The red line in both panels corresponds to a correlation of 1.

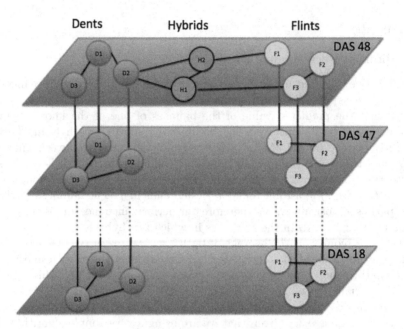

**Fig. 6.** Time line network. Three schematic layers, networks from DAS 18, 47 and 48, represent the full network of all 33 layers. Note that only on DAS 48 (and 28, not shown) hybrid nodes are present because biomass measurements are only available on those days. Therefore a prediction starting on e.g. DAS 47 must travel through the DAS 48 layer. Connections between layers are here only modelled between the same genotypes.

maximum path length is only 0.4 (remember that this is for shortest paths between Dents and Flints of the founder layers, and therefore much shorter than the general case quoted above). In summary, our preliminary analysis with a simple algorithm shows encouraging results in the direction of predicting hybrid performance from inbred phenotypic traits by graph-theoretic methods applied to the network proposed above.

## 5    Conclusion and Future Work

In this paper, we present a modelling framework for a large dataset of phenotypic features on a set of inbred and hybrid lines that are linked by a breeding function. We also present a first graph-theoretic algorithm that exploits correlation properties to predict hybrid performance. In the following we will outline how the proposed network and algorithm could be improved and extended.

Firstly, the algorithm could be improved in a number of ways. For example, it would be desirable to average over several paths that have a similar path length, or one could explore the network neighbourhood of the hybrid(s) in the path. For most of these extensions, suitable helper algorithms already exists and it is only a matter of implementing them in the given context. Another topic of interest is the relationship between the set of phenotypic traits used to compute the correlation between genotypes and the performance measure. Here we used all the 141 available phenotypic traits, which are somewhat redundant and in some cases uncorrelated to biomass. We are in the process of testing several selection methods to identify an optimal set of features.

The network $\mathcal{N}_t$ itself could also be improved and enriched. It is important to note that several other correlation structures that are related to different biological levels give rise to networks like $\mathcal{N}_t$. Combining several data sources (e.g. SNPs, RNA-seq and metabolic profiles) in a single network with the same nodes would generate networks with multi-weighted edges which could be analyzed by specific algorithms that consider also the reliability of information coming from each biological level. Moreover, additional information related to genotypes can be stored in the nodes, such as several performance measures that could all contribute to the final prediction.

Finally, a timeline network that collapses all networks $\mathcal{N}_t$ into a single network $M$ is of high interest. To that end, edges between nodes from different time points need to be added, again based on correlations. The simplest way to achieve this would be to add edges between the same genotypes at adjacent time points; however, one could also include all pairwise correlations between members of the same pool and even allow edges to jump several DASs. The structure of such a network is illustrated in Fig. 6. In the regime of network $M$, the same algorithms could be employed to predict final biomass from nodes in earlier time layers. Translated into the world of maize breeders, it might allow selecting promising hybrids at earlier developmental stages which would results in great savings in time and money. With this happy prospect in mind, our network model bears considerable potential for hybrid breeding.

# References

1. Andorf, S., Gärtner, T., Steinfath, M., Witucka-Wall, H., Altmann, T., Repsilber, D.: Towards systems biology of heterosis: a hypothesis about molecular network structure applied for the Arabidopsis metabolome. EURASIP J. Bioinform. Syst. Biol. **2009**(1), 1–12 (2009)
2. Breiman, L.: Random forests. Mach. Learn. **45**(1), 5–32 (2001)
3. Chen, D., Neumann, K., Friedel, S., Kilian, B., Chen, M., Altmann, T., Klukas, C.: Dissecting the phenotypic components of crop plant growth and drought responses based on high-throughput image analysis. Plant Cell **26**(12), 4636–4655 (2014)
4. Feher, K., Lisec, J., Römisch-Margl, L., Selbig, J., Gierl, A., Piepho, H.P., Nikoloski, Z., Willmitzer, L.: Deducing hybrid performance from parental metabolic profiles of young primary roots of maize by using a multivariate diallel approach. PLoS ONE **9**(1), e85435 (2014)
5. Gärtner, T., Steinfath, M., Andorf, S., Lisec, J., Meyer, R.C., Altmann, T., Willmitzer, L., Selbig, J.: Improved heterosis prediction by combining information on DNA- and metabolic markers. PLoS ONE **4**(4), e5220–547 (2009)
6. Groszmann, M., Greaves, I.K., Fujimoto, R., Peacock, W.J., Dennis, E.S.: The role of epigenetics in hybrid vigour. Trends Genet. **29**(12), 684–690 (2013)

7. Hastie, T., Tibshirani, R., Friedman, J.: The Elements of Statistical Learning. Springer Series in Statistics. Springer New York Inc., New York (2001)
8. Junker, A., Murayam, M.M., Weigelt-Fischer, K., Arana-Ceballos, F., Klukas, C., Melchinger, A.E., Meyer, R.C., Riewe, D., Altmann, T.: Optimizing experimental procedures for quantitative evaluation of crop plant performance in high throughput phenotyping systems. Frontiers in¡/CHECK¿. Front. Plant Sci. **5**, 770 (2015)
9. Klukas, C., Chen, D., Pape, J.M.: Integrated analysis platform: an open-source information system for high-throughput plant phenotyping. Plant Physiol. **165**(2), 506–518 (2014)
10. Klukas, C., Pape, J.M., Entzian, A.: Analysis of high-throughput plant image data with the information system IAP. J. Integr. Bioinform. **9**(2), 191 (2012)
11. Liaw, A., Wiener, M.: Classification and Regression by randomForest. R News **2**(3), 18–22 (2002)
12. Marbach, D., Costello, J.C., Küffner, R., Vega, N.M., Prill, R.J., Camacho, D.M., Allison, K.R., Aderhold, A., The DREAM5 Consortium, Kellis, M., Collins, J.J., Stolovitzky, G.: Wisdom of crowds for robust gene network inference. Nat. Methods **9**(8), 796–804 (2012)
13. Neumann, K., Klukas, C., Friedel, S., Rischbeck, P., Chen, D., Entzian, A., Stein, N., Graner, A., Kilian, B.: Dissecting spatio-temporal biomass accumulation in barley under different water regimes using high-throughput image analysis. Plant Cell and Environment, February 2015
14. Ogutu, J.O., Piepho, H.P.: Regularized group regression methods for genomic prediction: Bridge, MCP, SCAD, group bridge, group lasso, sparse group lasso, group MCP and group SCAD. BMC Proc. **8**(Suppl 5), S7 (2014)
15. Schnable, P.S., Springer, N.M.: Progress toward understanding heterosis in crop plants. Annu. Rev. Plant Biol. **64**, 71–88 (2013)
16. Steinfath, M., Gärtner, T., Lisec, J., Meyer, R.C., Altmann, T., Willmitzer, L., Selbig, J.: Prediction of hybrid biomass in Arabidopsis thaliana by selected parental SNP and metabolic markers. Theoret. Appl. Genet. **120**(2), 239–247 (2010)
17. Strogatz, S.H.: Exploring complex networks. Nature **410**(6825), 268–276 (2001)
18. Wold, H.: Soft Modelling By Latent Variables. Academic Press, London (1975)
19. Xu, S., Zhu, D., Zhang, Q.: Predicting hybrid performance in rice using genomic best linear unbiased prediction. PNAS **111**(34), 12456–12461 (2014)

# Automated Motion Analysis of Adherent Cells in Monolayer Culture

Zhen Zhang[1], Matthew Bedder[2], Stephen L. Smith[1(✉)],
Dawn Walker[3], Saqib Shabir[4], and Jennifer Southgate[4]

[1] Department of Electronics, University of York,
Heslington, York YO10 5DD, UK
{zz636, stephen.smith}@york.ac.uk
[2] Department of Computer Science, University of York,
Heslington, York YO10 5GW, UK
mb708@york.ac.uk
[3] Department of Computer Science and Insigneo Institute
for in Silico Medicine, University of Sheffield, Sheffield S1 4DP, UK
d.c.walker@sheffield.ac.uk
[4] Jack Birch Unit, Department of Biology, University of York,
Heslington, York YO10 5DD, UK
jennifer.southgate@york.ac.uk

**Abstract.** This paper presents a novel method for tracking and characterizing adherent cells in monolayer culture. A system of cell tracking employing computer vision techniques was applied to time-lapse videos of replicate normal human uro-epithelial cell cultures exposed to different concentrations of adenosine triphosphate (ATP), acquired over a 20 h period. Subsequent analysis, comprising feature extraction, demonstrated the ability of the technique to successfully separate the modulated classes of cell.

## 1 Introduction

The bladder is lined by urothelium, a remarkable tissue that forms the tightest and most efficient self-repairing barrier in the body. After physical or other damage, the urothelium switches rapidly and transiently from a stable mitotically-quiescent barrier into a highly regenerative state. The mechanisms involved in this switch are poorly understood, but are central to understanding the pathophysiology of the human urinary bladder.

The urothelium is reported to respond to mechanical and chemical stimulation by releasing various transient mediators, including adenosine triphosphate (ATP), which have been proposed to play a role in mediating neuronal signalling [1]. In addition, the urothelium expresses purinergic P2X and P2Y receptors and channels that are responsive to ATP released from autocrine or paracrine sources [2]. The outcome of such signalling is incompletely understood, as it could have a feedback role in modulating neuronal signalling, but alternatively could play a more direct role in urothelial barrier repair [2]. It has been further suggested that aberant expression of receptors

© Springer International Publishing Switzerland 2015
M. Lones et al. (Eds.): IPCAT 2015, LNCS 9303, pp. 185–194, 2015.
DOI: 10.1007/978-3-319-23108-2_16

and/or mediator release by the urothelium is involved in dysfunctional diseases of the bladder, including idiopathic detrusor instability and interstitial cystitis [3, 4].

Despite the literature reporting expression of these channels and receptors by the urothelium, consensus has been confounded by contradictions in experimental approaches, including the species, specificity of reagents, and the nature of the tissue preparation (reviewed [5]). Our approach to address these questions has been to develop a cell and tissue culture system for investigating normal human urothelial cells and tissues in vitro. In our work, we have shown that stimulation of P2 receptors with exogenous ATP enhanced scratch wound repair, as did the addition of the ecto-ATPase inhibitor ARL-67156, which prevents the breakdown of autocrine-produced ATP. By contrast, blockade of P2 activity inhibited scratch wound repair in either the presence or absence of ATP [2]. This indicates that ATP is one of the major factors released upon damage and contributes to the regenerative phenotype.

To understand the effect of ATP on urothelial cell phenotype, time-lapse videos have been generated of urothelial cell cultures to which exogenous ATP has been applied. This paper describes the development of an automated method for objective measurement of these videos using computer vision techniques, followed by the extraction of features, with the aim of describing the cell behaviour. These processes are described in detail in Sect. 2, where measurements have been obtained from the set of videos of urothelial cell cultures with and without exogenous ATP added. The results with statistical analysis are presented in Sect. 3, verified by manual analysis.

## 2 Methods

The automated analysis of cell motion comprises the following sequence of analysis: capturing images of the cells at regular intervals using videomicroscopy, tracking cells within the video on a frame-by-frame basis using custom-written software, followed by characterisation of cell movement through the extraction of specially designed features. Each of these stages is considered in further detail in the following sections.

### 2.1 Cell Culture and Videomicroscopy

Normal human urothelial (NHU) cells were established in culture as finite (non-immortalised) cell lines and maintained as detailed elsewhere [6]. For ATP experiments, cells were seeded in a 12 well plate and the growth medium was supplemented with 0, 10 or 50 μM ATP in replicates of four. Cultures were observed by differential interference contrast videomicroscopy (Olympus IX81 microscope) in an environmental chamber with an automated mechanical stage. Timelapse videos were compiled from individual images captured digitally every 10 min over a 20 h time period. A sample frame from one such video is illustrated in Fig. 1.

### 2.2 Cell Tracking

Custom-written software was developed to undertake automated cell tracking using the OpenCV computer vision programming library [7]. In order to track the relative

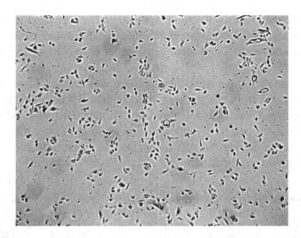

**Fig. 1.** Sample frame from timelapse video of NHU cells in culture.

movement of cells within a video, each frame undergoes processing to identify the likely locations of cells. This process takes the raw videos as an input, performs common preprocessing to each frame, and then either tries to identify the likely location of cells, or track the location of previously located cells.

Each video frame initially undergoes Gaussian blurring to remove noise, followed by simple thresholding against a predetermined fixed value, resulting in a binary image separating the foreground and background (i.e. the cells from the frame background). A distance transform is then applied to this binary image, resulting in frames where the centre of large cells (or masses of cells) have a larger value, the edge of cells have a low value, and the background has a value of 0.

In order to efficiently estimate the locations of the centres of cells, the local maxima of the distance images are computed. Local maxima are then selected from the highest to lowest scoring, with a small area around each selected maxima being filtered out to reduce the number of selections made within the body of a cell. The (x,y) coordinates of the selected maxima are then used as estimations of the locations of cells within the frame.

To estimate the location of a cell within a frame, given the location within the previous frame, the distance image around the previous cell location is first multiplied by a simple Gaussian filter. The maximal pixel value in this region can then be used to estimate the new cell location. This approach, although simplistic, is demonstrated to be effective. The usage of the distance image promotes matches with the centre of cells, whilst the application of the Gaussian filter means that matches are preferred that are close to the original location of the cell.

Although the process for tracking cells works well over the videos so far tested, it is unable to consistently identify and track cell locations for the duration of the videos. In order to detect as many cells as possible, a large number of potential cell locations are initially calculated, with many of these quickly converging into the same locations. Similarly, the cell tracking process can occasionally fail to track the location of cells within frames, meaning that if cell detection were only to be performed on the first frame of the video then many cells would not be tracked in the latter parts of the video. These difficulties associated with tracking cells can be due to cell proliferation - giving rise to new cells, cell death – the loss of cells, and cells moving in and out of the field of view.

In order to cope with these issues, an approach was adopted where duplicated cell locations are removed from the tracking process and cell detection is performed at regular intervals to find new candidate locations. This approach is found to be effective and results in location data for a sufficient number of cells over the duration of the video to adequately describe the cell population behavior. The entire cell tracking process is summarised in Fig. 2.

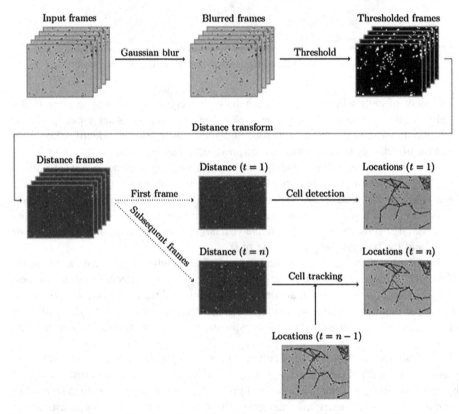

**Fig. 2.** The process used for detecting cell locations within a video, and tracking detected cells between video frames

### 2.3    Feature Extraction

Once the location of cells has been identified for each frame of the video in the form of (x,y) coordinate pairs, it is possible to extract features with the aim of describing the cell population behaviour. This was undertaken using the MATLAB programming environment [8] and to illustrate the processing applied, a single cell from a video analysed is taken as an example to demonstrate how features of interest are calculated. The cell selected has been tracked from a video of NHU cell culture with 50 μM exogenous ATP. As this cell was successfully tracked from the beginning to the end of the video, its path can be shown graphically as depicted in Fig. 3.

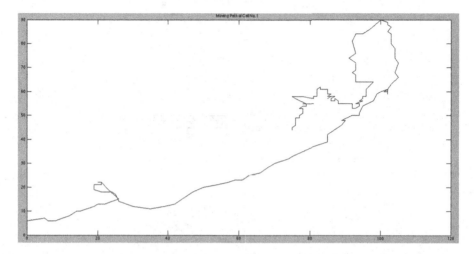

**Fig. 3.** Example tracking of a single cell over a 20 h video sampled every 10 min.

## 2.4    Choice of Features

The choice of features to extract from the videos was made on the basis of their subsequent use in describing cell behaviour. For this reason features of (i) cell migration speed and (ii) migration persistence were defined as described below.

### 2.4.1    Cell Migration Speed

Speed of an object is the rate of change of its position. In this case, the aim is to obtain the migration speed of a cell from a video, which can be determined by calculating the number of pixels travelled over a certain time interval. The time interval applicable in this context is that between two consecutive video frames, at a frame rate of one every 10 min. The migration speed is therefore simply obtained by calculating the Euclidean distance between the two pairs of coordinates for the cell between consecutive frames. This is shown graphically in Fig. 4 where the initial position of the cell is at coordinates (74,32) and in the subsequent frame, coordinates (75,33). Hence, the distance travelled by this cell over time $dt$ (10 min), and subsequently, its speed, can be calculated. The migration speed of all cells tracked during the entire video was calculated in the same way.

### 2.4.2    Cell Migration Persistence

In cell migration, persistence is one of the features in which biologists are most interested and can be described as the tendency of cells to change direction. Hence, obtaining the direction of travel of the cell in each frame of the video is essential for calculating migration persistence.

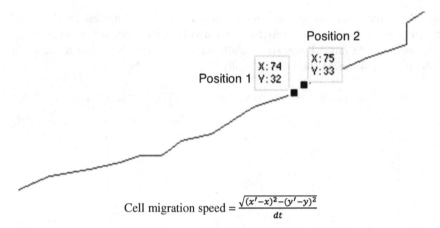

$$\text{Cell migration speed} = \frac{\sqrt{(x'-x)^2-(y'-y)^2}}{dt}$$

**Fig. 4.** Example calculation of cell migration speed.

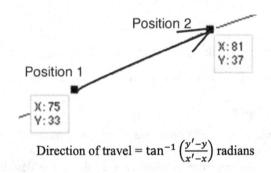

$$\text{Direction of travel} = \tan^{-1}\left(\frac{y'-y}{x'-x}\right) \text{ radians}$$

**Fig. 5.** Direction of travel of cell migration.

Figure 5 shows how the angle of the vector formed from the coordinates of the cell in consecutive frames of the video can be used to determine the direction of travel. Angular Velocity is defined here as the rate of change of the direction of travel of a cell over subsequent frames. Figure 6 illustrates an example calculation over two consecutive frames.

## 3  Results

Four time-lapse videos have been generated from three classes of NHU cell cultures: (i) a control culture with no ATP; (ii) a culture with 10 μM ATP; and (iii) a culture with 50 μM ATP. The average cell migration speeds and average angular velocity for each video is presented in Table 1.

By applying analysis of variance (ANOVA), it can be seen in Fig. 7 that the separation between the three classes for migration speed are statistically significant.

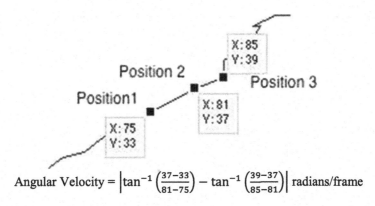

$$\text{Angular Velocity} = \left| \tan^{-1}\left(\frac{37-33}{81-75}\right) - \tan^{-1}\left(\frac{39-37}{85-81}\right) \right| \text{ radians/frame}$$

**Fig. 6.** Example calculation of cell migration persistence over two consecutive frames.

**Table 1.** Automated average migration speed and average angular velocity values for a control culture with no ATP, a culture with 10 µM ATP and a culture with 50 µM ATP.

| Cell culture video | Average migration speed (pixels/frame) | Average angular velocity (rads/frame) |
|---|---|---|
| Control 1 | 3.52 | 1.31 |
| Control 2 | 3.75 | 1.35 |
| Control 3 | 3.45 | 1.37 |
| Control 4 | 3.56 | 1.36 |
| 10 uM ATP1 | 3.04 | 1.52 |
| 10 uM ATP2 | 3.09 | 1.39 |
| 10 uM ATP3 | 3.08 | 1.52 |
| 10 uM ATP4 | 3.01 | 1.46 |
| 50 uM ATP1 | 2.00 | 1.79 |
| 50 uM ATP2 | 2.22 | 1.74 |
| 50 uM ATP3 | 2.06 | 1.77 |
| 50 uM ATP4 | 1.83 | 1.85 |

Verification of these results was confirmed by comparing with manual tracking of 15 random cells for each experimental condition as shown in Fig. 7. Similarly, results for angular velocity, shown in Fig. 8, also demonstrate good separation between the three sets of culture conditions.

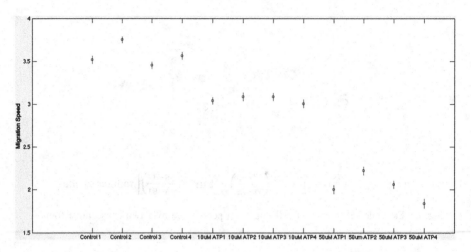

**Fig. 7.** Automated calculation of cell migration persistence. Average migration speeds are shown in F-distribution form for Control, 10 uM ATP and 50 uM ATP Videos: small circles mark the mean of the group and the bars the 95 % confidence interval.

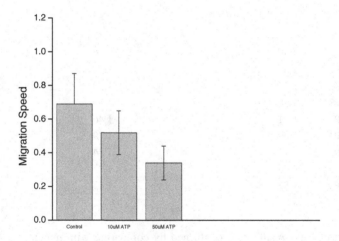

**Fig. 8.** Calculation of cell migration persistence using manual tracking.

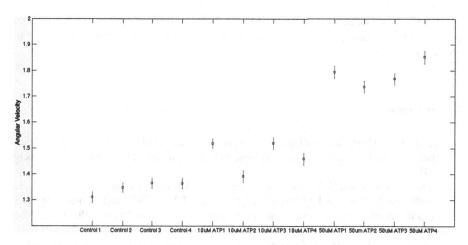

**Fig. 9.** Automated calculation of cell angular velocity. Average migration speeds are shown in F-distribution form for Control, 10 uM ATP and 50 uM ATP Videos: small circles mark the mean of the group and the bars represent 95 % confidence intervals.

## 4  Conclusion and Discussion

The application of an automated approach for tracking and characterizing adherent cells in monolayer culture has been presented that compensates for the loss and gain of cells during the course of the video – a commonly encountered problem in tracking systems of this type. This opens the possibility of lineage-tracking within mixed populations, where the automated approach removes the labour intensive nature of manual tracking. Here, the classification of cells from cultures with and without exogenous ATP has identified statistically significant effects on cell behaviour that will contribute to understanding of urothelial tissue repair mechanisms and the role of ATP (Fig. 9).

We do not yet understand the implications of the results for urothelial biology as we have not previously had the capability to measure these aspects of cell behaviour. Purinergic receptor activation by ATP as a result of cellular release upon injury has been implicated in cellular migration during the restitution of cornea [9], airway [10, 11] and bladder [2] epithelial tissues, suggesting common effects. The underpinning mechanisms are poorly understood, although amplification of TGFβ1-induced actin remodelling associated with cell migration has been suggested [11]. The results of our study were unexpected in that we found ATP to give increased angular velocity and reduced migration speeds. However, we studied the effect of exogenous application of ATP rather than local endogenous ATP release and although speculative, it may be that the precise mode (concentration/locality) of ATP stimulation might affect cell migration response including directionality, for recruitment to the wound.

Current and future work is focused on further characterisation of the behaviour of cells using additional extracted features that relate to *cohesivity* – the tendency of cells to form and stay in clumps. The nature of this contact can be described in terms of the duration of the contact and number of cells that form a clump, as well as the effect on post-contact migration speed and angular velocity of individual cells. These characteristics relate to the physical nature of the contact that forms between cells and provides critical information as to what extent they are transient or more sustained in character.

# References

1. Birder, L.A.: Urothelial signaling. In: Andersson, K.-E., Michel, M.C. (eds.) Urinary Tract. Handbook of Experimental Pharmacology, vol. 202, pp. 207–231. Springer, Heidelberg (2011)

2. Shabir, S., Cross, W., Kirkwood, L.A., Pearson, J.F., Appleby, P.A., Walker, D., Eardley, I., Southgate, J.: Functional expression of purinergic P2 receptors and transient receptor potential channels by the human urothelium. Am. J. Physiol. Ren. Physiol. **305**, F396–F406 (2013)

3. Birder, L.A., de Groat, W.C.: Mechanisms of disease: involvement of the urothelium in bladder dysfunction. Nature clinical practice. Urology **4**, 46–54 (2007)

4. Sun, Y., Chai, T.C.: Up-regulation of P2X3 receptor during stretch of bladder urothelial cells from patients with interstitial cystitis. J. Urol. **171**, 448–452 (2004)

5. Yu, W., Hill, W.G.: Defining protein expression in the urothelium: a problem of more than transitional interest. Am. J. Physiol. Ren. Physiol. **301**, F932–F942 (2011)

6. Southgate, J., Hutton, K.A., Thomas, D.F., Trejdosiewicz, L.K.: Normal human urothelial cells in vitro: proliferation and induction of stratification. Lab. Invest. J. Tech. Methods Pathol. **71**, 583–594 (1994)

7. Bradski, G.: The opencv library. Doctor Dobbs J. **25**, 120–126 (2000)

8. MATLAB R2014A: The MathWorks, Inc., Natick, Massachusetts, United States

9. Boucher, I., Rich, C., Lee, A., Marcincin, M., Trinkaus-Randall, V.: The P2Y2 receptor mediates the epithelial injury response and cell migration. Am. J. Physiol. Cell Physiol. **299**, C411–C421 (2010)

10. Wesley, U.V., Bove, P.F., Hristova, M., McCarthy, S., van der Vliet, A.: Airway epithelial cell migration and wound repair by ATP-mediated activation of dual oxidase 1. J. Biol. Chem. **282**, 3213–3220 (2007)

11. Takai, E., Tsukimoto, M., Harada, H., Sawada, K., Moriyama, Y., Kojima, S.: Autocrine regulation of TGF-β1-induced cell migration by exocytosis of ATP and activation of P2 receptors in human lung cancer cells. J. Cell Sci. **125**, 5051–5060 (2012)

# Neural Modelling
# and Neural Networks

# Community Detection as Pattern Restoration by Attractor Neural-Network Dynamics

Hiroshi Okamoto[1,2(✉)]

[1] Research & Development Group,
Fuji Xerox Co., Ltd., Hadano, Kanagawa, Japan
hiroshi.okamoto@fujixerox.co.jp
[2] RIKEN Brain Science Institute, Saitama, Japan

**Abstract.** Densely connected parts in networks are referred to as "communities". Community structure is a hallmark of a variety of real-world networks; individual communities form functional modules constituting complex systems described by networks. Therefore, revealing community structure in networks is essential to approaching and understanding complex systems described by networks. This is the reason why network science has made a great deal of effort to develop effective and efficient methods for detecting communities in networks. Here we examine a novel type of community detection, which has not been examined so far but will be of great practical use. Suppose that we are given a set of source nodes that includes some (but not all) of "true" members of a particular community; suppose also that the set includes some nodes that are not the members of this community (i.e., "false" members of the community). We propose to detect the community from this "imperfect" and "inaccurate" set of source nodes using attractor neural-network dynamics. Community detection achieved by the proposed method can be viewed as restoration of the original pattern from a deteriorated pattern, which is also analogous to cue-triggered recall of short-term memory in the brain. We demonstrate the effectiveness of the proposed method using synthetic networks and real social networks for which correct communities are known.

**Keywords:** Complex network · Community · Local detection · Pattern restoration · Cell assembly · Short-term memory

## 1 Introduction

In network science, a group of nodes in a network that are densely connected within this group and are less densely connected with nodes outside the group is referred to as a "community". Community structure is a fundamental property of a variety of biological, social and engineering networks. Development of effective and efficient algorithms to detect communities in networks has been a big challenge of network science in the last decade [1, 2].

Many of algorithms proposed up to now aim to exhaustively detect all the communities in a given network. Another, more economic approach is to detect only a community to which given source nodes belongs. Starting from source nodes, one explores the network; exploration will continue until certain criteria are met. The explored region

© Springer International Publishing Switzerland 2015
M. Lones et al. (Eds.): IPCAT 2015, LNCS 9303, pp. 197–207, 2015.
DOI: 10.1007/978-3-319-23108-2_17

or a part of it is then judged as a community to which source nodes belongs. This type of community detection is described as "local" because it requires only knowledge about the structure of a local part of the network around the community to be detected.

Several algorithms for local community detection have already been proposed [3–9]. Nevertheless, most of them are designed to detect a community to which a 'single' source node belongs. However, we often encounter practical situations where community detection for a set of source nodes is required. For instance, suppose that we know several members of a particular community and wish to find all the members belonging to this community. Let $S$ be a set of these known members; namely, $S$ expresses our "imperfect" knowledge about this community. Our knowledge might also be "inaccurate" and therefore $S$ might include some "false" members (i.e. members that do not belong to this community). Our task is to find all the members that "truly" belong to this community starting from the imperfect and inaccurate set of members. Most of the algorithms proposed up to now [3–9] are unable to efficiently accomplish this kind of local community detection.

Recently the present author has proposed a method of local community detection by attractor neural-network dynamics for a single source node [10]. Here we extend this method so as to make it applicable to local community detection for a set of source nodes that might be imperfect and inaccurate. We demonstrate the effectiveness of the proposed method using synthetic networks and real social networks for which correct communities are known.

We can consider individual communities in a given network as correct patterns embedded in this network and a set of source nodes, which might include false as well as true members of a particular community, as a pattern deteriorated from the original pattern. Detection of the correct community from this imperfect and inaccurate set of source nodes can therefore be viewed as restoration of the original pattern from the deteriorated pattern. This is analogue to pattern completion by Hopfield's attractor neural-network dynamics [11].

In fact, we have devised the proposed method inspired by possible neural mechanisms of cue-triggered recall of short-term memory in the brain. Hence we additionally discuss biological relevance of local detection of communities in networks.

## 2    Methods

### 2.1    Neural-Network Dynamics

Let $N$ be the number of nodes of a network from which we wish to detect communities. Let $\mathbf{A} = (A_{nm})$ ($n, m = 1, \cdots, N$) be the adjacency matrix of this network, where $A_{nm}$ is the weight of the link from node $m$ to node $n$. For simplicity this study deals with networks with undirected links, where the adjacency matrix is symmetric ($A_{nm} = A_{mn}$), but our discussion can easily be extend to community detection from directed networks.

Now we compare individual nodes to neurons and individual links to synaptic connections between neurons. Let $p_n(t)$ and $f_n(t)$ be the "potential" and the "activity" of neuron $n$ at time $t$, respectively. We assume that the relationship between $p_n(t)$ and

$f_n(t)$ is given by a threshold-linear function (Fig. 1), which models the relationship between the membrane potential and the firing rate of pyramidal cells [12]:

$$f_n(t) = \Theta(p_n(t) - \theta)p_n(t),\qquad(1)$$

where $\Theta(x) = 0$ for $x < 0$ and $\Theta(x) = 1$ for $x \geq 0$.

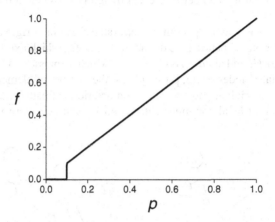

**Fig. 1.** Threshold-linear relationship between the potential $p$ and the activity $f$ defined by Eq. (1) for $\theta = 0.1$.

Time evolution of potentials of individual neurons is described by the dynamics defined by

$$p_n(t) = \sum_{m=1}^{N} T_{nm} f_m(t-1) + \frac{f_n(t-1)}{F(t-1)}[F_0 - F(t-1)] \qquad (F_0 > 0),\qquad(2)$$

where $T_{nm} \equiv A_{nm}/\sum_{n'=1}^{N} A_{n'm}$ and $F(t) \equiv \sum_{n=1}^{N} f_n(t)$. The first term on the right-hand side describes propagation of activities from neurons making synaptic connections onto neuron $n$ to neuron $n$. The second term models competition between neurons for a finite resource $F_0 - F$; such competition, which is generally considered to occur owing to activation of inhibitory interneurons, is common in cortical network architecture [13].

One can easily verify that the sum of the potential of all neurons is kept constant with time

$$\sum_{n=1}^{N} p_n(t) = F_0.\qquad(3)$$

This property is important as it stabilizes the neural-network dynamics (2), keeping it from falling into pathological states such as flare-up or extinction of activities of all neurons. Without loss of generality one can set $F_0 = 1$.

## 2.2 Imperfect and Inaccurate Set of Source Nodes: Deteriorated Pattern

Let $C$ be the correct community that we wish to detect from a given network (indicated by black nodes in the right of Fig. 2); $S$ be a set of source nodes (indicated by blue and red nodes in the left of Fig. 2); $T$ be the set of nodes in $S$ that belong to $C$ ("true" members indicated by blue nodes in the left of Fig. 2); $F$ be the set of nodes in $S$ that never belong to $C$ ("false" members indicated by red nodes in the left of Fig. 2). Thus, $S = T \cup F$.

Now we can view $S$ as a "pattern" deteriorated from the original pattern, $C$. To characterize a deteriorated pattern, we define $p_S \equiv |S|/|C|$, which expresses the level of partialness of $S$ to $C$, and define $r_F \equiv |F|/|C|$, which expresses to what extent $S$ is contaminated by false nodes. Here, $|X|$ stands for the number of elements of set $X$. Our task is to restore the original pattern, $C$, from a deteriorated pattern, $S$. Figure 2 gives intuitive illustration of local community detection viewed as pattern restoration.

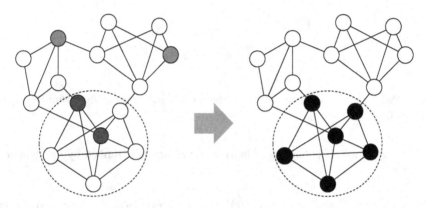

**Fig. 2.** Local detection of a community as pattern restoration. The subnetwork enclosed by dashed line expresses the community (the original pattern) to be detected. Filled nodes (either blue or red) in the left are source nodes, with blue and red nodes being "true" and "false" members of the community, respectively. In the right, the members of the detected community (restored pattern) are expressed by black nodes (Color figure online).

## 2.3 Local Community Detection

Given a set of source nodes, $S$, local community detection as pattern restoration is performed as follows. First we set the initial condition

$$
\begin{aligned}
p_n(0) &= 1/|S| \quad \text{if node } n \in S; \\
p_n(0) &= 0 \quad\quad \text{otherwise.}
\end{aligned}
\tag{4}
$$

Note that this initial condition satisfies normalization, $\sum_{n=1}^{N} p_n(0) = 1$. Activities of neurons spread from the source nodes and then propagate in the network according to Eq. (2). Iterative calculation of Eq. (2) eventually leads to the steady-state distribution of potentials, $\left\{ p_1^{(\text{stead})}, \cdots, p_N^{(\text{stead})} \right\}$. This steady-state distribution defines the community detected for the deteriorated pattern, $S$. The $p_n^{(\text{stead})}$ has a graded value ranging from 0 to 1, which expresses the level of belongingness of node $n$ to the detected community.

To quantify to what extent the detected community accurately restores the correct community (original pattern), we use the mean average precision (MAP), which is a metrics widely used for evaluating goodness of document retrieval with ranking. Let $K$ be the total number of correct communities in the network. For each correct community $C_k$, a set of source nodes $S_k$ is generated, and then the steady-state distribution of activities of nodes are calculated as described above. Then, all the nodes are sorted in descending order of their steady-state activities. The average precision for detection of community $k$ is defined by

$$\text{AP}_k = \frac{1}{|C_k|} \sum_{i=1}^{N} \frac{z_i}{i} \left( 1 + \sum_{j=1}^{i-1} z_j \right). \tag{5}$$

Here, $z_i = 1$ if the node with the $i$-the largest steady-state potential is a member of community $C_k$; $z_i = 0$ otherwise. The average precision is larger for more accurate restoration of correct communities. It is maximized ($\text{AP}_k = 1$) when the top $|C_k|$ nodes are all the members of community.

The MAP is the mean of the average precision over the communities:

$$\text{MAP} = \frac{1}{K} \sum_{k=1}^{K} \text{AP}_k. \tag{6}$$

Specifically, MAP = 1 means that the local detection of communities (namely, restoration of patterns) is perfect.

## 2.4  State-of-the-Art Study

Other than the proposed method, spreading activation [14], also known as personalized PageRank algorithm (PPR) [15], is an efficient and probably unique method for detecting a set of nodes that are densely connected and are relevant to a given set of source nodes. We choose PPR as a competitor to the proposed method for a state-of-the art study.

Spreading of activities by PPR is given by

$$p_n(t) = (1 - \rho) \sum_{m=1}^{N} T_{nm} f_m(t - 1) + \rho b_n \qquad (0 \leq \rho \leq 1) \tag{7}$$

where

$$\begin{aligned} b_n &= 1/|S| \quad \text{if node } n \in S; \\ b_n &= 0 \qquad \text{otherwise.} \end{aligned} \tag{8}$$

The first term on the right-hand side of (7) describes that activities propagate in the network along links with probability $1 - \rho$; the second term describes that an activity at any node warps to source nodes with probability $\rho$ and hence serves as a bias towards source nodes. Thus, in parallel to spreading of activities along links, activities are constantly pushed to $S$; eventually activities are localized to $C$ in the steady state.

### 2.5  Biological Relevance

In fact, we have devised the above method of local community detection inspired by possible neural mechanisms of short-term memory recall in the brain. The cell assembly hypothesis, which is prevailing in neuroscience, states that neurons coding the same item tend to be mutually connected [16], thus forming a densely connected group, referred to as a "cell assembly". Short-term memory recall of a particular item is associated with sustained activation of the cell assembly coding this item [17, 18]. Reverberative propagation of neuronal activities in the cell assembly is the mechanism of sustainment of activation of neurons constituting the cell assembly [19].

The brain stores numerous memory items. Different items are coded by different cell assemblies. This means that the neural-network dynamics underlying short-term memory recall is multi-stable [11, 20]. Initial activation of a fraction of neurons, which serves as a cue, determines which stable state (attractor) will be selected. These mechanisms achieve cue-triggered short-term memory recall of a particular item.

It is quite natural to compare cell assemblies in brain networks to communities in real-world networks; a fraction of neurons that are initially activated serving as a cue to a set of source nodes. Thus, our method of local community detection by use of the neural-network dynamics as given by Eq. (2) models cue-triggered short-term memory recall in the brain.

## 3  Results

### 3.1  Local Detection of Communities from Synthetic Benchmark Networks

First we evaluated the performance of local community detection as pattern restoration by neural-network dynamics using synthetic benchmark networks for which we knew correct communities. A method for synthesizing networks with community structure has been proposed by Lancichinetti et al. [21, 22]. The number of communities and their sizes can be controlled by adjusting the parameter values. We synthesized a network of $N = 1000$ nodes with $K = 30$ communities. The parameter values used for synthesizing the network and the statistics of communities are given in Appendix.

For each of the 30 communities, a set of source nodes was generated. In the first experiment, $p_S$ was fixed at the maximum value ($p_S = 1.0$) while $r_F$ was ranged from 0.1 to 1.0. The $p_S = 1.0$ means that the number of source nodes was approximately the same as the number of members of the correct community to be detected. The fractional value of $r_F$ expresses the ratio of the number of false nodes in the set of source nodes to the number of true members of the community to be detected, which models the level of inaccuracy of our knowledge about the community. The MAP is shown as a function of $r_F$ in Fig. 3a (open circles). The MAP takes near maximum values ($\sim 1.0$) up to considerably high values of $r_F$; even for $r_F = 0.8$, where only $\sim 20\%$ of source nodes are "true" and the remaining $\sim 80\%$ are "false", the MAP is considerably high ($> 0.8$).

In the next experiment, we ranged $p_S$ from 0.1 to 1.0 while keeping $r_F = 0.3$. The fractional value of $p_S$ expressed the ratio of source nodes to the total number of true members of the community and models the imperfectness of our knowledge about the community to be detected. For smaller $p_S$, the correct community must be restored from its smaller portion. Figure 3b (open circles) shows the MAP as a function of $p_S$. The MAP takes near maxim values for a wide range of the value of $p_S$. Even for

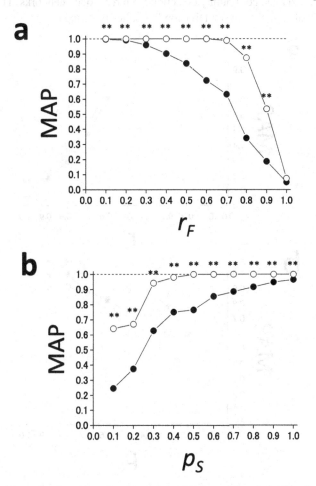

**Fig. 3. a**, MAP as a function of $r_F$ for $p_S = 1.0$. **b**, MAP as a function of $p_S$ for $r_F = 0.3$. The dashed line in each diagram indicates the maximum of possible MAP values. Open and filled circles indicate the results obtained by the proposed method and PPR, respectively. Double asterisks (**) above open circles indicate the statistical significance of the superiority of the proposed method to PPR ($p < 0.01$, paired $t$-test).

$p_S = 0.3$, where only $\sim$ 30 % of nodes of the community to be detected were given as source nodes with $\sim$ 30 % of them being "false", a considerably high value of MAP (> 0.9) was obtained.

The results obtained by PPR are also shown in Fig. 3 (filled circles). The MAP values obtained by the proposed method always exceed or are equal to those obtained by PPR ($p < 0.01$, paired $t$-test). Thus, local detection of communities from synthetic benchmark networks by the proposed method outperforms that by PPR.

## 3.2   Local Detection of Communities from Real Social Networks

Next we examined local community detection from real social networks. The network of American football games between Division I-A colleges during regular season Fall 2000,

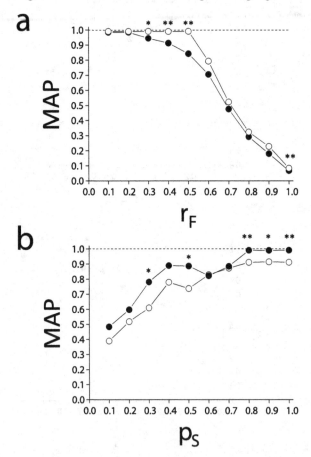

**Fig. 4. a**, MAP as a function of $r_F$ for $p_S = 1.0$. **b**, MAP as a function of $p_S$ for $r_F = 0.3$. The dashed line in each diagram indicates the maximum of possible MAP values. Open and filled circles indicate the results obtained by the proposed method and PPR, respectively. A single asterisk (*) and double asterisks (**) above open circles indicate the statistical significance of the superiority of the proposed method to PPR (* for $p < 0.05$ and ** for $p < 0.01$, paired $t$-test).

as compiled by Girvan and Newman (2002) [23], has 115 nodes with each corresponding to a team. Each team (except for five independent teams) belongs to one of 12 conferences. Since teams belonging to the same conference have more games with each other than with those belonging to different conferences, teams belonging to the same conference will tend to form a community.

The MAP was examined as a function of either $r_F$ or $p_S$. Essentially the same results as was obtained for synthetic networks was gained; the MAP takes near maximum values for a wide range of the values for $r_F$ (Fig. 4a, open circles) or $p_S$ (Fig. 4b, open circles).

Filled circles in Fig. 4a and b indicate the results obtained by PPR. The MAP values obtained by the proposed method are always above or equal to those obtained by PPR. Since the superiority of the proposed method to PPR was undetected for $p_S = 0.4$, 0.6 and 0.7 in Fig. 4b by paired $t$-test, we looked into individual $AP_k$'s (see Eq. (5)) obtained by the both methods. We found that for these values of $p_S$, most of $AP_k$'s obtained by the proposed method was maximum ($AP_k = 1$) but a few $AP_k$'s were near zero; all $AP_k$'s obtained by PPR were less than 1 but their distribution was much more moderate. The extreme separation of $AP_k$ values obtained by the proposed method produced a specious large variation, which masked statistical significance of the superiority of the proposed method. Results obtained above suggest that local detection of communities from real social networks by the proposed method outperforms that by PPR.

## 4 Discussion

Most algorithms for local community detection proposed up to now are intended to detect the community to which a single source node belongs [3–10]. In the present study we have examined a different type of local community detection task, which is difficult to efficiently achieve for the previously proposed algorithms [3–9]. This task requires local detection of a correct community given a set of source nodes that includes false as well as true members of this community. Such a set of source nodes expresses our imperfect and inaccurate knowledge about the community to be detected. We have proposed to perform this type of local community detection by use of neural-network dynamics. Local community detection in this way can be viewed as restoration of the original pattern (namely, the correct community to be detected) from a deteriorated pattern (an imperfect and inaccurate set of nodes) and is analogue to pattern completion by Hopfield's attractor neural-network dynamics [11].

We have demonstrated the effectiveness of the proposed method using synthetic benchmark networks and real social networks for which correct communities are known. The proposed method accurately restores the correct communities and outperforms PPR, the only existing method that achieves detection of the densely connected part around the source nodes.

To our surprise, local community detection from an imperfect and inaccurate set of source nodes as demonstrated in the present study has been little examined in literature of network science. However, we believe that this type of local community detection is of great practical use. For instance, just given only partial and inaccurate knowledge about the members of a particular community, one can extract all the members of this community from social networks by use of the method proposed here. This will enable effective advertising or recommendation of products or services focused on this community.

We have devised the proposed method of local community detection inspired by neural mechanisms of cue-triggered short-term memory retrieval in the brain. The present study exemplifies that modelling real brain mechanisms is beneficial for creating new information processing architecture.

**Acknowledgments.** This study was partly supported by KAKENHI (15K00418).

## Appendix: Synthetic Benchmark Network

The benchmark network used in Sect. 3.1 was synthesized using the software downloaded from [15] under the following settings: Number of nodes 1000; average degree 15; maximum degree 50; exponent for the degree distribution 2; exponent for the community size distribution 1; mixing parameter 0.2; minimum for the community sizes 20; maximum for the community sizes 50. The synthesized network has 30 communities, with the following size occurrences: (size, occurrence) = (20, 1), (21, 2), (23, 1), (26, 4), (27, 2), (28, 2), (29, 1), (30, 1), (31, 2), (33, 1), (34, 1), (36, 1), (37, 1), (38, 1), (40, 1), (41, 3), (42, 1), (44, 2), (471, 1), (62, 1).

## References

1. Fortunato, S.: Community detection in graphs. Phys. Rep. **486**, 75–174 (2010)
2. Newman, M.E.J.: Communities, modules and large-scale structure in networks. Nature Phys. **8**, 25–31 (2012)
3. Bagrow, J.P., Bollt, E.M.: Local method for detecting communities. Phys. Rev. E **72**, 046108 (2005)
4. Clauset, A.: Finding local community structure in networks. Phys. Rev. E **72**, 026132 (2005)
5. Luo, F., Wang, J., Promislow, E.: Exploring local community structures in large networks. Web Intell. Agent Syst. **6**, 387–400 (2008)
6. Chen, J., Zäiane, O., Goebel, R.: Local community identification in social networks. In: International Conference on Advances in Social Network Analysis and Mining (ASONAM2009), pp. 237–242 (2009)
7. Lancichinetti, A., Fortunato, S., Kertesz, J.: Detecting the overlapping and hierarchical community structure in complex networks. New J. Phys. **11**, 033015 (2009)
8. Branting, L.K.: Context-sensitive detection of local community structure. Soc. Netw. Anal. Min. **2**, 279–289 (2012)
9. Chen, Q., Wu, T.-T., Fang, M.: Detecting local community structures in complex networks based on local degree central nodes. Physica A **392**, 529–537 (2013)
10. Okamoto, H.: Local detection of communities by neural-network dynamics. In: Mladenov, V., Koprinkova-Hristova, P., Palm, G., Villa, A.E., Appollini, B., Kasabov, N. (eds.) ICANN 2013. LNCS, vol. 8131, pp. 50–57. Springer, Heidelberg (2013)

11. Hopfield, J.J.: Neural networks and physical systems with emergent collective computational abilities. Proc. Nat. Acad. Sci. USA **79**, 2554–2558 (1982)
12. Tuckwell, H.: Introduction to Theoretical Neurobiology: Volume 2 Nonlinear And Stochastic Theories. Cambridge University Press, Cambridge (1988)
13. Rabinovich, N.I., Volkovskii, A., Lecanda, P., Heurta, R., Abarbanel, H.D., Laurent, G.: Phys. Rev. Lett. **87**, 068102 (2001)
14. Collins, A.M., Loftus, E.F.: Spreading-activation theory of semantic processing. Psychol. Rev. **82**, 407–428 (1975)
15. Page, L. et al.: The PageRank Citation Ranking: Bringing Order to the Web. Technical report, Stanford InfoLab (1998). ilpubs.stanford.edu:8090/422/
16. Hebb, D.O.: Organization of Behaviour. Wiley, New York (1949)
17. Funahashi, S., Bruce, C.J., Goldman-Rakic, P.S.: Mnemonic coding of visual space in the monkey's dorsolateral prefrontal cortex. J. Neurophysiol. **61**, 331–349 (1989)
18. Churchland, A.K., Kiani, R., Shadlen, M.N.: Decision making with multiple alternatives. Nat. Neurosci. **11**, 693–702 (2008)
19. Durstewitz, D., Seamans, J.K., Sejnowski, T.J.: Neurocomputational model of working memory. Nature Neuroscience (suppl. 3), pp. 1184–1191 (2000)
20. Wang, X.-J.: Neural dynamics and circuit mechanisms of decision-making. Curr. Opin. Neurobiol. **22**, 1–8 (2012)
21. Lancichinetti, A., Fortunato, S., Radicchi, F.: Benchmark graphs for testing community detection algorithms. Phys. Rev. E **78**, 046110 (2008)
22. http://santo.fortunato.googlepages.com/benchmark.tgz
23. Girvan, M., Newman, M.E.J.: Community structure in social and biological networks. Proc. Nat. Acad. Sci. USA **99**, 7821–7826 (2002)

# Feature Learning HyperNEAT: Evolving Neural Networks to Extract Features for Classification of Maritime Satellite Imagery

Phillip Verbancsics[(✉)] and Josh Harguess

SPAWAR Systems Center Pacific, 53560 Hull Street, San Diego, CA, USA
{phillip.verbancsics,joshua.harguess}@navy.mil

**Abstract.** Imagery analysis represents a significant aspect of maritime domain awareness; however, the amount of imagery is exceeding human capability to process. Unfortunately, the maritime domain presents unique challenges for machine learning to automate such analysis. Indeed, when object recognition algorithms observe real-world data, they face hurdles not present in experimental situations. Imagery from such domains suffers from degradation, have limited examples, and vary greatly in format. These limitations are present satellite imagery because of the associated constraints in expense and capability. To this end, the Hypercube-based NeuroEvolution of Augmenting Topologies approach is investigated in addressing some such challenges for classifying maritime vessels from satellite imagery. Results show that HyperNEAT learns features from such imagery that allows better classification than Principal Component Analysis (PCA). Furthermore, HyperNEAT enables a unique capability to scale image sizes through the indirect encoding.

**Keywords:** Image classification · Artificial neural networks · Hyperneat

## 1 Introduction

A proliferation of remote and unmanned platforms has resulted in an exponential growth of streaming imagery. This increase in data presents many challenges. For example, the growth in data from unmanned systems has resulted in an unexpected paradox: Unmanned systems require more labor to operate than manned systems, that is, to remotely operate a single Reaper unmanned aerial vehicle (UAV) requires more than 80 analysts during a single mission [9]. Thus cost in human labor is increasing despite the extensive deployment of "unmanned" systems because of the challenges in analyzing the data. In particular, a significant reason for the labor costs of unmanned systems is the need to analyze streaming visual data. The challenge in addressing this overload of visual data is replicating the flexibility and capability of humans on the required tasks; therefore, approaches that can replicate the power of the human visual system are of significant interest.

© Springer International Publishing Switzerland 2015
M. Lones et al. (Eds.): IPCAT 2015, LNCS 9303, pp. 208–220, 2015.
DOI: 10.1007/978-3-319-23108-2_18

Successfully designing an automated approach to such visual tasks often result in significant labor costs. First, articulating how the visual system works is difficult. That is, it is easy for an analyst to recognize an object, but it is more difficult to describe the method by which that recognition occurs, therefore experts must invest significant time to design such systems. Second, data is constantly changing with new objects to be recognized or existing objects changing appearance that requires system redesign. Thus, machine learning is employed to automate this design of imagery analysis. Indeed, artificial neural networks are resurgent thanks to breakthroughs in deep learning that have led to state-of-the-art results in a number of challenging domains [2].

In particular, deep learning approaches have achieved remarkable performance in a number of object recognition benchmarks, often achieving the current best performance on these tasks. Such object recognition tasks where deep learning has achieved the best results include the MNIST hand-written digit dataset [17,21], traffic sign recognition [5], and the ImageNet Large-Scale Visual Recognition Challenge [18]. Real-world conditions, however, can degrade the effectiveness of such approaches [13]. Imagery from the maritime domain (the focus of this paper) presents barriers to learning in the form of small data sets and many image formats, in addition to occlusions, distortions, and degradation [19]. In addition, processing maritime imagery presents limitations in the form of resources, such as bandwidth or processing power, that necessitates dynamic resolution changes to adapt to time constraints, that is, imagery may be transmitted or processed at lower resolutions when required by constraints.

This paper explores the Hypercube-based NeuroEvolution of Augmenting Topologies (HyperNEAT; [11,12,24]) in the domain of vessel classification from satellite imagery. HyperNEAT has shown promise in simple visual discrimination tasks [6,10,16,33] and in extracting features for handwritten digits [27,30,31] but its potential has yet to be fully explored in real-world imagery. HyperNEAT is applied to a vessel-classification problem that has been investigated with traditional computer vision techniques [14,15,19,20] by learning feature extractors for this image classification task and is compared against Principal Component Analysis (PCA). In addition, the robustness of HyperNEAT's learning to variations in how the data is normalized is examined and robustness to changes in image resolution for a trained solution is explored. Results show that Hyper-NEAT is capable of creating feature extractors that outperform PCA extracted features in classification performed by $k$-Nearest Neighbors (KNN). Furthermore, HyperNEAT can effectively learn with different variations on normalizing the input data. Interestingly, these different normalizations do impact the types of features learned and can aid in overcoming challenges in the data set (e.g. biases towards a particular class). Finally, HyperNEAT's indirect encoding allows the resizing of the feature extractor, thus training at one resolution can be applied to any resolution.

(a)              (b)              (c)              (d)

**Fig. 1.** BCCT200 Data Set Examples. Example images from each class of the BCCT200 data set: (a) barge, (b) cargo vessel, (c) container vessel, and (d) tanker.

## 2    Background

The satellite imagery data set being examined and the neuro-evolutionary methods that underlie the approach are described in this section.

### 2.1    BCCT200

Automatic vessel classification is an important goal for many military and security-related applications. Towards that goal, Harguess et al. [14] created the Barge, Cargo, Container, and Tanker (BCCT200) dataset. The data originates from image chips (sub-images within a larger image containing desired objects) created by the RAPid Image Exploitation Resource (RAPIER®), developed by Space and Naval Warfare (SPAWAR) Systems Center Pacific (SSC Pacific) [3], which provides automatic vessel detection in overhead satellite imagery in support of image analysts. The BCCT200 dataset was created by first hand-labeling the image chips into the appropriate vessel type category to form an database of 4 classes with 200 images in each class. Then the images were rotated, cropped, and resized as described in [14,15]. In this paper, the BCCT200-resize is the dataset for performance comparisons and is referred to generically as BCCT200. Examples from each class of the BCCT200 dataset can be seen in Fig. 1.

### 2.2    NeuroEvolution of Augmenting Topologies (NEAT)

The NeuroEvolution of Augmenting Topologies (NEAT) algorithm [26] is a popular neuroevolutionary approach that has been proven in a variety of challenging tasks, including particle physics [1,35], simulated car racing [4], RoboCup Keepaway [28], function approximation [34], and real-time agent evolution [23], among others [26]. NEAT starts with a population of small, simple ANNs that increase their complexity over generations by adding new nodes and connections through mutation. That way, the topology of the network does not need to be known a priori; NEAT searches through increasingly complex networks as it evolves their connection weights to find a suitable level of complexity. The techniques that facilitate evolving a population of diverse and increasingly complex networks are described in detail in Stanley and Miikkulainen [26]; the important concept for the approach in this paper is that NEAT is an evolutionary method that discovers the right topology and weights of a network to maximize performance on a task. The next section reviews the extension of NEAT called HyperNEAT that allows it to effectively train large neural structures.

## 2.3 CPPNs and HyperNEAT

Hypercube-based NEAT (HyperNEAT; [12,24]) is an extension of NEAT that enables effective evolution of high-dimensional ANNs through indirect encoding. The effectiveness of the geometry-based learning in HyperNEAT has been demonstrated in domains such as multi-agent predator prey [7,8] and RoboCup Keepaway [32]. A full description of HyperNEAT is in Stanley et al. [24].

The main idea in HyperNEAT is that geometric relationships are learned though an indirect encoding that describes how the *weights* of the ANN can be *generated* as a function of geometry. Unlike a direct representation, wherein every connection in the ANN is described individually, an indirect representation describes a pattern of parameters without explicitly enumerating each such parameter. That is, information is reused in such an encoding, which is a major focus in the field of GDS from which HyperNEAT originates [25,29]. Such information reuse allows indirect encoding to search a compressed space. HyperNEAT discovers the *regularities* in the geometry and learns from them.

The indirect encoding in HyperNEAT is called a *compositional pattern producing network* (CPPN; [22]), which encodes the *weight pattern* of an ANN [11,24]. The idea behind CPPNs is that geometric patterns can be encoded by a *composition of functions* that are chosen to represent common regularities. Consider a CPPN that takes four inputs labeled $x_1$, $y_1$, $x_2$, and $y_2$; this point in four-dimensional space can *also* denote the connection between the two-dimensional points $(x_1, y_1)$ and $(x_2, y_2)$. The output of the CPPN for that input thereby represents the weight of that connection (Fig. 2). By querying every pair of points, the CPPN can produce an ANN, wherein each queried point is the position of a neuron. Because the connection weights are produced as a function of their endpoints, the pattern is produced with *knowledge* of the domain geometry.

In summary, HyperNEAT evolves the topology and weights of the CPPN that *encodes* ANN weight patterns. An extension of HyperNEAT called HyperNEAT with Link Expression Output (HyperNEAT-LEO) was introduced to constrain connectivity with a bias towards modularity [33]. This extension separates the decision of weight magnitude and expression into *two* different CPPN outputs and seeds the LEO with the concept of locality. The next section introduces the approach in this paper that applies HyperNEAT to deep learning.

## 3 Feature Learning HyperNEAT

Although the HyperNEAT method succeeds in a number of challenging tasks [11,12,24,32] by exploiting geometric regularities, it is just beginning to be

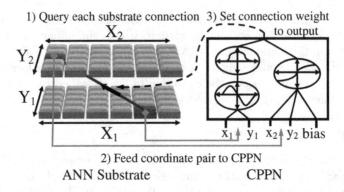

**Fig. 2. A CPPN Describes Connectivity.** A grid of nodes, called the ANN *substrate*, is assigned coordinates. (1) Every connection between layers in the substrate is queried by the CPPN to determine its weight; the line connecting layers in the substrate represents a sample such connection. (2) For each such query, the CPPN inputs the coordinates of the two endpoints. (3) The weight between them is output by the CPPN. Thus, CPPNs can generate regular patterns of connections.

applied to visual domains [27,30,31]. Because HyperNEAT learns as a function of domain geometry, it is well suited towards such domains where geometric relationships are important. Conventional HyperNEAT trains a CPPN that defines an ANN that is the solution, that is, the produced ANN is applied directly to the task and then the ANN's performance on that task determines the CPPN's fitness score. However, Feature Learning HyperNEAT trains an ANN that transforms inputs into features based upon domain geometry and then the features are given to another machine learning approach to solve the problem. Thus, the performance of this learned solution then defines the fitness score of the CPPN for HyperNEAT (Fig. 3). In this way, HyperNEAT acts as a reinforcement learning approach that determines the best features to extract for another machine learning approach to maximize performance on the task. The experiments exploring Feature Learning HyperNEAT in maritime vessel classification are detailed in the next section.

## 4   Experimental Setup

These investigations are conducted on the BCCT200 data set, which is a real-world data from commercial satellite imagery consisting of 200 images each of barge, cargo, container, and tanker vessels. BCCT200 images are 150×300 pixels, but for these experiments the images are scaled down to 28 × 28 pixels. The goal for machine learning is to correctly classify vessel contained within each image. The data set is split into three sub-sets: Training (100 images/class), Evaluation (50 images/class), and Testing (50 images/class). During evolution, features are extracted from the training set to train KNN ($k = 3$), features extracted from the evaluation set determine the performance of KNN on unseen

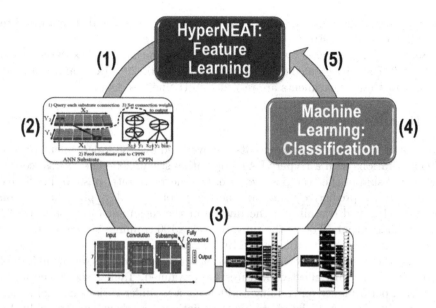

**Fig. 3. Feature Learning HyperNEAT.** To learn features, HyperNEAT trains CPPNs (1) that generate the connectivity for a defined ANN substrate (2). The ANN substrate processes the inputs from a data set to produce a set of features (3). These features are given to another machine learning algorithm (4) that learns to perform the task (e.g. image classification). Machine learning then produces a solution that is evaluated on testing data. The performance of the solution on data not seen during training provides the fitness score of the CPPN for HyperNEAT (5). In this way, HyperNEAT learns better features for the machine learning approach to perform its task.

data and inform the fitness of the feature extractor for HyperNEAT, and after evolution is completed, the testing set is applied to determine performance on data not seen during evolution. For both HyperNEAT and PCA, the number of features extracted is set to 48. To compare with PCA, the ANN substrate is constrained to no hidden layers, ensuring the features extracted are linear as are the features extracted by PCA. Thus HyperNEAT's ANN substrate is to a $28 \times 28$ pixels input to 48 outputs with no hidden layers and sigmoid activation function on the outputs.

A number of methods exist for pre-processing pixels prior to presentation, chief among them statistical normalization approaches. This paper explores the effect of varying three different normalization settings on the performance of feature learning with HyperNEAT. (1) Normalization is varied between max normalization (simply dividing all pixels by the max possible value), mean normalization, and normalizing to the standard deviation. (2) The data from which the normalization is calculated is varied between all the pixels from all the images, all the pixels within a single image, and the pixels at a particular location. (3) The range of the values is set to either unipolar ($[0, 1]$) or bipolar ($[-1, 1]$). The normalization values are calculated through the training set and then applied to the evaluation and testing sets.

Scaling is implemented by applying a solution trained at the $28 \times 28$ resolution to the BCCT200 data scaled down from $150 \times 300$ to $20 \times 20$, $50 \times 50$, and $100 \times 100$. Results for these experiments are recorded in the next section.

## 5   Results

For each of these experiments, results are averaged over 30 independent runs of 1500 generations with a HyperNEAT population size of 200. The fitness score is a weighted sum of the true positive rate, true negative rate, positive predictive value, negative predictive value, and accuracy for each class plus the fraction correctly classified overall and the inverse of the mean square error from the correct label outputs. For all the runs, the same 100 training, 50 evaluation, and 50 testing images are given from the BCCT200 data set.

First, the effects of different input normalization techniques on the ability of HyperNEAT to learn effective features are examined. Table 1 shows the mean champion performance on training and testing sets under various normalization approaches (Note: Associated standard deviations are shown in Appendix A). Mean classification performance on the training data ranges from a minimum of 0.843 with mean normalization per pixel and bipolar range to 0.896 with standard deviation normalization per pixel with bipolar range. For performance on the testing data, the minimum performance is 0.659 with mean normalization per pixel with bipolar range and maximum performance is 0.787 with standard deviation normalization per image with unipolar range. The best combined training and testing performance is with data normalized to the standard deviation per image and unipolar range (0.857 training, 0.787 testing). Interestingly, only the mean per pixel with bipolar range normalization approach significantly ($p < 0.01$) underperforms the rest. Additionally, the peak performers for the training data and testing data both have greater performance than other approaches with significance ($p < 0.01$).

Complementing the mean results, Table 2 shows the results of the champion under each normalization approach with the best overall performance, that is, it classifies the combined training and testing data better than the other champions. PCA provides the baseline performance comparison with mean normalized performance of 0.855 training and 0.753 testing and standard deviation normalized performance of 0.863 training and 0.733 testing. Peak HyperNEAT classification performance in on the training data ranges from a minimum of 0.858 with mean normalization of all images and bipolar range to 0.920 with standard deviation normalization of all images with unipolar range. For performance on the testing data, the minimum performance is 0.693 with mean normalization per pixel with bipolar range and maximum performance is 0.803 with max normalization with bipolar range. The best combined training and testing performance is with data normalized to the max and bipolar range (0.893 training, 0.803 testing). Note that the best peak performers (max bipolar, and, standard deviation of all images with unipolar range) do not come from the normalization

**Table 1.** Mean training and testing classification performance by normalization approach.

| | | Normalizer | | | | | |
|---|---|---|---|---|---|---|---|
| | | Mean | Dev. | Mean Bipolar | Dev. Bipolar | Max | Max Bipolar |
| | All Images | 0.85 \| 0.71 | 0.87 \| 0.72 | 0.86 \| 0.73 | 0.85 \| 0.71 | 0.86 \| 0.71 | 0.88 \| 0.74 |
| Data | Per Image | 0.85 \| 0.70 | *0.86 \| 0.79* | 0.87 \| 0.72 | 0.86 \| 0.71 | | |
| | Per Pixel | 0.85 \| 0.71 | 0.85 \| 0.70 | 0.84 \| 0.66 | **0.90** \| 0.73 | | |

**Table 2.** Peak training and testing classification performance of feature learning HyperNEAT by normalization approach versus PCA.

| | | Normalizer | | | | | |
|---|---|---|---|---|---|---|---|
| | | Mean | Dev. | Mean Bipolar | Dev. Bipolar | Max | Max Bipolar |
| | All Images | 0.88 \| 0.75 | **0.92** \| 0.74 | 0.86 \| 0.76 | 0.86 \| 0.72 | 0.88 \| 0.75 | *0.89 \| 0.80* |
| Data | Per Image | 0.88 \| 0.73 | 0.88 \| 0.77 | 0.90 \| 0.75 | 0.86 \| 0.75 | | |
| | Per Pixel | 0.88 \| 0.72 | 0.87 \| 0.70 | 0.87 \| 0.69 | 0.91 \| 0.75 | | |
| | PCA | 0.86 \| 0.75 | 0.86 \| 0.73 | | | | |

approaches the achieved the best mean champion performance on those data sets. Furthermore, all HyperNEAT extracts better features for the training set under all normalization approaches except the mean of all images with bipolar range. However, HyperNEAT only exceeds PCA's testing performance under three normalization approaches; Standard deviation per image with unipolar range, Max with bipolar range, and Mean of all images with bipolar range.

Further comparison between PCA and HyperNEAT features can be seen in the performance per image class from the BCCT200 data set from peak performers learned by HyperNEAT. Features extracted by PCA are strongly biased towards correctly classifying the barge class, correctly specifying 98 % of barges as barges. However, such features are weak at classifying cargo vessels, only specifying 50 % correctly as cargo. Most of the normalization approaches result in HyperNEAT learning features with similar biases to PCA. Figure 4a shows that the best peak performer (max bipolar normalization) classifies 96 % of barges correctly and only 56 % of cargo vessels correctly. Interestingly, one normalization approach, mean normalization per image with bipolar range, finds a peak performer that begins to overcome this bias in the data (Fig. 4b). In this case, HyperNEAT learns features that has KNN classifying barges with 92 % correct and cargo with 72 % correct.

Exploring this contrast in features learned, Fig. 5 shows the classification performance on the training set over time per class and overall for HyperNEAT with max bipolar and mean per image bipolar normalization, respectively. The max bipolar normalization quickly converges to a set of features on the training set that allows 95 % of barges, 89 % of tankers, 85 % of container vessels, 81 % of cargo vessels to be correctly classified. Once it reaches this level, very little is gained from further learning. In contrast, the mean per image with bipolar range

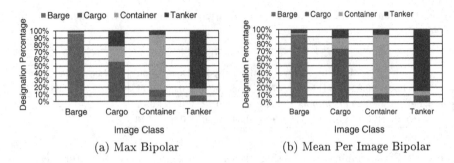

(a) Max Bipolar     (b) Mean Per Image Bipolar

**Fig. 4. Confusion of Image Classes with Feature Learning HyperNEAT.** Shown is the percentage of instances (vertical axis) each class of image (horizontal axis) is designated as a particular class by KNN when given features extracted by the peak performer (champion with best combined training and testing performance across 30 runs) learned by HyperNEAT with max normalization with bipolar range (a) and mean per image normalization with bipolar range (b).

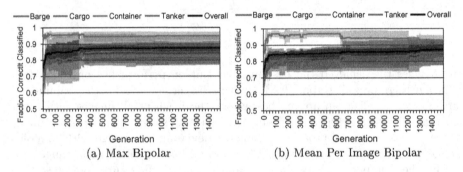

(a) Max Bipolar     (b) Mean Per Image Bipolar

**Fig. 5. Classification Performance by Class Over Evolution.** Shown is the mean of each generation champion's fraction correctly classified for each image class max normalization with bipolar range (a) and mean per image normalization with bipolar range (b) from the training set.

normalization starts by finding features biased towards barges, with classification rates of 97 % for barges, 82 % for container vessels, and 80 % each for tankers and cargo vessels, but then learns features that are more balanced for classifying all the classes, with classification rates of 93 % (barges), 86 % (cargo, container), and 85 % (tankers).

Finally, scaling is examined by taking the best performer from the mean per image normalization with bipolar range and applying the trained CPPN to larger $(50 \times 50, 100 \times 100)$ and smaller $(20 \times 20)$ ANN substrates. These different sized ANN substrates are then applied to the BCCT200 images scaled down from $150 \times 300$ pixels to the appropriate resolution to extract features without further training. As a comparison, Fig. 6 shows the performance of PCA extracted features when PCA is re-run at each resolution. PCA's features improve performance on the training set as the resolution increases, with 85 % correct

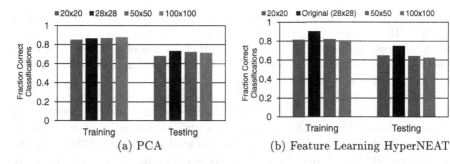

**Fig. 6. Classification Performance through Scaling.** Shown is the performance of PCA and Feature Learning HyperNEAT extracted features at $20 \times 20$, $28 \times 28$, $50 \times 50$ and $100 \times 100$ resolutions. Note that PCA is retrained at each resolution and Feature Learning HyperNEAT scales without further training.

classification at $20 \times 20$, 86 % at $28 \times 28$, 87 % at $50 \times 50$, and 88 % at $100 \times 100$. Interestingly, the same improvement in performance is not found in the testing set performance, where PCA's feature extraction initially improves from 68 % correct classification at $20 \times 20$ to 73 % at $28 \times 28$, but then declines to 72 % at $50 \times 50$ and 71 % at $100 \times 100$. This decline in testing performance demonstrates a two-fold challenge with such real-world data sets, in that small sample amounts prevent generalization especially when given more information per data sample.

Scaling for Feature Learning HyperNEAT is tested by taking the single best performer from the mean per image normalized with a bipolar range runs. Unlike PCA, the learned CPPN is not re-trained at each resolution, instead it is trained at a single resolution ($28 \times 28$) and then the is CPPN applied to smaller ($20 \times 20$) and larger ($50 \times 50$, $100 \times 100$) substrates to extract features from the different resolution images. In this way, learning can be transferred from one resolution to another without costly re-training. HyperNEAT's correct classification performance at the trained resolution ($28 \times 28$) is 90 % for training data and 75 % for testing data. Performance on both training and testing data is degraded when scaling to the different resolutions with training and testing performance being 81 % and 65 %, 82 % and 64 %, and, 81 % and 63 % for the 20, 50, and 100 scales, respectively. Note again that these are results with *no* further training on the new resolutions, but are simply scaled from the original trained solution. These results further contrast from prior results in scaling with traditional computer vision approaches [19], where performance drops by more than half under similar scale changes.

## 6   Discussion and Conclusion

Performance of image classification algorithms often depends on the type and quality of the data on which the algorithm is used. Imagery gathered under real-world conditions suffer from problems often not observed in experimental conditions, such lack of data samples, degraded quality, and different formats or resolutions. In particular, satellite imagery encounters these problems frequently, with imagery of high or low resolution, noise from the sensor or transmission process, and be blocked by clouds or other occlusions. Overall, algorithms deployed to the real-world are more likely to face less than ideal data, thus it essential for successful designs to handle such data.

Prior work with HyperNEAT has shown promise in vision tasks that operate directly on raw pixels. This paper extended such work by investigating Feature Learning HyperNEAT in the challenging classification of maritime of maritime vessels from satellite imagery represented by the BCCT200 data set. Results showed that HyperNEAT discovers superior linear feature extractors versus PCA under different manipulations of the data. Furthermore, the correct pre-processing allows HyperNEAT to overcome a strong bias present that can be present in small data sets. Finally, HyperNEAT demonstrates an ability extend what has been learned to different image resolutions, which is challenging to other machine learning approaches. Thus, HyperNEAT presents a unique approach to feature learning for imagery that can enable capabilities that are not present or difficult in current computer vision approaches.

**Acknowledgments.** This work was supported and funded by the SSC Pacific Naval Innovative Science and Engineering (NISE) Program.

## A    Result Standard Deviations

**Table 3.** Standard deviation of training and testing classification performance by normalization approach.

| | | Normalizer | | | | | |
|---|---|---|---|---|---|---|---|
| | | Mean | Dev. | Mean Bipolar | Dev. Bipolar | Max | Max Bipolar |
| Data | All Images | 0.023 \| 0.039 | 0.031 \| 0.021 | 0.009 \| 0.021 | 0.014 \| 0.027 | 0.019 \| 0.024 | 0.017 \| 0.029 |
| | Per Image | 0.019 \| 0.023 | *0.019 \| 0.042* | 0.024 \| 0.035 | 0.014 \| 0.041 | | |
| | Per Pixel | 0.027 \| 0.016 | 0.021 \| 0.022 | 0.015 \| 0.037 | **0.044** \| 0.05 | | |

## References

1. Aaltonen, T., et al.: Measurement of the top quark mass with dilepton events selected using neuroevolution at CDF. Phys. Rev. Lett. **102**, 152001–152008 (2009)
2. Bengio, Y., Courville, A., Vincent, P.: Representation learning: a review and new perspectives. IEEE Trans. Pattern Anal. Mach. Intell. **35**(8), 1798–1828 (2013)
3. Buck, H., Sharghi, E., Guilas, C., Stastny, J., Morgart, W., Schalcosky, B., Pifko, K.: Enhanced ship detection from overhead imagery. In: SPIE Defense and Security Symposium, vol. 6945. International Society for Optics and Photonics (2008)

4. Cardamone, L., Loiacono, D., Lanzi, P.L.: On-line neuroevolution applied to the open racing car simulator. In: Proceedings of the IEEE CEC., Piscataway, NJ, USA. IEEE Press (2009)
5. Ciresan, D., Meier, U., Masci, J., Schmidhuber, J.: Multi-column deep neural network for traffic sign classification. Neural Netw. **32**, 333–338 (2012)
6. Coleman, O.J.: Evolving neural networks for visual processing. Ph.D. thesis, The University of New South Wales, Kensingtion, Austrailia (2010)
7. D'Ambroiso, D., Stanley, K.O.: Evolving policy geometry for scalable multiagent learning. In: Proceedings of the 9th AAMAS, New York, NY, USA, p. 8. ACM Press (2010)
8. D'Ambrosio, D.B., Stanley, K.O.: Generative encoding for multiagent learning. In: Proceedings of GECCO 2008, New York, NY. ACM Press (2008)
9. Eggers, J.: Plenary talk. In: ICCRTS (2013)
10. Gauci, J., Stanley, K.O.: Generating large-scale neural networks through discovering geometric regularities. In: Proceedings of GECCO 2007, New York, NY, p. 8. ACM (2007)
11. Gauci, J., Stanley, K.O.: A case study on the critical role of geometric regularity in machine learning. In: Proceedings of the 23rd AAAI Conference, Menlo Park, CA. AAAI Press (2008)
12. Gauci, J., Stanley, K.O.: Autonomous evolution of topographic regularities in artificial neural networks. Neural Comput. **22**(7), 1860–1898 (2010)
13. Hall, D., McCool, C., Dayoub, F., Sunderhauf, N., Upcroft, B.: Evaluation of features for leaf classification in challenging conditions. In: IEEE WACV, Waikola, HI, January 2015
14. Harguess, J., Parameswaran, S., Rainey, K., Stastny, J.: Vessel classification in overhead satellite imagery using learned dictionaries. In: SPIE Optical Engineering+ Applications, p. 84992F–84992F. International Society for Optics and Photonics (2012)
15. Harguess, J., Rainey, K.: Are face recognition methods useful for classifying ships? In: IEEE AIPR Workshop, 2011, pp. 1–7. IEEE (2011)
16. Hausknecht, M., Khandelwal, P., Miikkulainen, R., Stone, P.: Hyperneat-gpp: A hyperneat-based atari general game player. In: Proceedings of GECCO 2012, Philadelphia, Pennsylvania, p. 8. ACM Press, July 2012
17. Hinton, G.E., Osindero, S., Teh, Y.-W.: A fast learning algorithm for deep belief nets. Neural Comput. **18**, 1527–1554 (2006)
18. Krizhevsky, A., Sutskever, I., Hinton, G.E.: Imagenet classfication with deep convolutional neural networks. In: Advances in NIPS (2012)
19. Rainey, K., Parameswaran, S., Harguess, J.: Maritime vessel recognition in degraded satellite imagery. In: Proceedings of SPIE Automatic Target Recognition XXIV, vol. 9090. International Society for Optics and Photonics, June 2014
20. Rainey, K., Stastny, J.: Object recognition in ocean imagery using feature selection and compressive sensing. In: IEEE AIPR Workshop, pp. 1–6. IEEE (2011)
21. Rifai, S., Dauphin, Y.N., Vincent, P., Bengio, Y., Muller, X.: The manifold tangent classifer. In: Advances in NIPS (2011)
22. Stanley, K.O.: Compositional pattern producing networks: a novel abstraction of development. Genet. Program. Evolvable Mach. Spec. Issue Dev. Syst. **8**(2), 131–162 (2007)
23. Stanley, K.O., Bryant, B.D., Miikkulainen, R.: Real-time neuroevolution in the NERO video game. IEEE Trans. Evol. Comput. Spec. Issue Evol. Comput. Games **9**(6), 653–668 (2005)

24. Stanley, K.O., D'Ambrosio, D.B., Gauci, J.: A hypercube-based indirect encoding for evolving large-scale neural networks. Artif. Life **15**, 185–212 (2009)
25. Stanley, K.O., Miikkulainen, R.: A taxonomy for artificial embryogeny. Artif. Life **9**(2), 93 (2003)
26. Stanley, K.O., Miikkulainen, R.: Competitive coevolution through evolutionary complexification. J. Artif. Intell. Res. **21**, 63 (2004)
27. Szerlip, P.A., Morse, G., Pugh, J.K., Stanley, K.O.: Unsupervised feature learning through divergent discriminative feature accumulation (2014). arXiv:1406.1833v2, abs/1406.1833
28. Taylor, M.E., Whiteson, S., Stone, P.: Comparing evolutionary and temporal difference methods in a reinforcement learning domain. In: Proceedings of GECCO 2006, New York, NY, pp. 1321–1328. ACM Press, July 2006
29. Turing, A.M.: The chemical basis of Morphogenesis. R. Soc. Lond. Philos. Trans. Ser. B **237**, 37–72 (1952)
30. Verbancsics, P., Harguess, J.: Deep learning through generative and developmental system. In: Proceedings of the Genetic and Evolutionary Computation (GECCO 2014) companion, p. 103. ACM (2014)
31. Verbancsics, P., Harguess, J.: Image classification using generative neuroevolution for deep learning. In: IEEE Winter Conference on Applications of Computer Vision (WACV), Waikola, HI, January 2015
32. Verbancsics, P., Stanley, K.O.: Evolving static representations for task transfer. J. Mach. Learn. Res. **11**, 1737–1769 (2010)
33. Verbancsics, P., Stanley, K.O.: Constraining connectivity to encourage modularity in hyperneat. In: Proceedings of GECCO 2011, New York, NY. ACM Press (2011)
34. Whiteson, S.: Improving reinforcement learning function approximators via neuroevolution. In Proceedings of the 4th AAMAS, New York, NY, USA, pp. 1386–1386. ACM (2005)
35. Whiteson, S., Whiteson, D.: Stochastic optimization for collision selection in high energy physics. In: IAAI 2007, Vancouver, British Columbia, Canada. AAAI Press, July 2007

# Improving Crossover of Neural Networks in Evolution Through Speciation

Phillip Verbancsics[✉]

SPAWAR Systems Center Pacific, 53560 Hull Street, San Diego, CA, USA
`phillip.verbancsics@navy.mil`

**Abstract.** Crossover is an important genetic operator that re-combines beneficial genes together and rapidly traverses the fitness landscape. Unfortunately, neuro-evolution (NE) has not experienced the benefits of crossover. Indeed, observations have shown that crossover has been detrimental to NE approaches. Tangentially, speciation has become an important feature in NE for diversity maintenance; however, such speciation research has focused on *what* measure is driving speciation versus *how* the measure determines species. This research posits that appropriate speciation implementations enable effective crossover by determining an individual's potential mating partners. Prior speciation research demonstrated the impact of restricting the mating pools of genomes on search performance. This paper investigates these concepts in the context of NE and results demonstrate; (1) the impact of speciation implementation in NE, (2) crossover's negative effect on search in NE, and (3) a novel speciation approach that enables effective crossover in NE.

**Keywords:** Speciation · Crossover · Artificial neural networks · NEAT

## 1 Introduction

Speciation and crossover are two prominent features of natural evolution that have significant mutual interaction [2]. In nature, speciation results in reproductive isolation that, in turn, limits the mating partners for sexual reproduction. This limitation of mating partners for reproduction then influences the genetic make-up of a population [1]. Diverging genetics among populations can be reinforced through crossover, which potentially results in speciation. Thus speciation plays an important role in crossover and vice versa. Similarly, evolutionary algorithms have explored speciation and crossover as important factors in evolutionary search, but not the interaction of the two aspects of evolution. This paper investigates the interaction of speciation and crossover and their impact on performance, in particular their effect on neuro-evolution (NE).

Interestingly, evolutionary approaches for artificial neural networks (ANNs) that rely on crossover have historically been shown to have diminished performance relative to approaches that emphasize mutation [14], despite evidence that natural evolution benefits from crossover [3]. This negative performance

© Springer International Publishing Switzerland 2015
M. Lones et al. (Eds.): IPCAT 2015, LNCS 9303, pp. 221–232, 2015.
DOI: 10.1007/978-3-319-23108-2_19

impact is thought to be due to the permutation problem, that is, multiple neural network genotypes can encode functionally equivalent phenotypes. This many-to-one mapping results in crossover for neural networks being ineffective. Additionally, the connectionist nature of neural networks presents a difficulties in representing the "building blocks" necessary for effective crossover. These limitations have resulted in dismissal of crossover as an operator in NE.

The effectiveness of crossover is influenced by the pool of available mating partners. Such pools are limited by reproductive isolation, which is reinforced by speciation [6]. Speciation has proven to be popular in evolutionary algorithms to encourage the formation of "niches" to preserve diversity, that is, rather than being the driver of limited mating pools, the focus is on preventing the population from prematurely converging [4]. Indeed, such speciation has proven effective in NE for preserving diversity in the population and improving performance [9]. Further research has focused on alternative measures to induce speciation, such as genotypic, phenotypic, and behavioral [8]. These alternate approaches are meant to preserve different types of diversity, thus research has focused creating "niches" rather than better mating pools for crossover.

In contrast, this paper explores how varying the heuristic that is applied to the speciation metric in order to select the species for genomes can influence. Four different heuristics for selecting species are investigated: First Compatible, Most Compatible, Parental, and Uncanny Valley (described in Sect. 3). The experiments to explore the interaction of speciation heuristic and crossover are performed in classic benchmark and calibration domains for NE (XOR and double-pole balancing) under high mutation and high crossover rates. Results from the experiments demonstrate the negative effect that relying highly on crossover has on performance, with the majority of speciation heuristics performing worse under such conditions. However, changing the heuristic results in different performance profiles. Indeed, Most Compatible is consistently the worst performing heuristic, while Uncanny Valley is the highest performer and performs similarly well under both high and low crossover conditions. These results indicate the importance of the effect of speciation heuristics on the performance of crossover and reveal the potential to unlock crossover's power through improved mating pool selection.

## 2    NeuroEvolution of Augmenting Topologies (NEAT)

This section briefly reviews the NEAT evolutionary algorithm [9,10], a prominent method that evolves ANNs. NEAT evolves connection weights as well as adds new nodes and connections over generations, thereby increasing solution complexity. It has been proven to be effective in challenging control and decision making tasks [10–13]. NEAT starts with a population of small, simple ANNs that increase their complexity over generations by adding new nodes and connections through mutation. That way, the topology of the network does not need to be known a priori; NEAT searches through increasingly complex networks as it evolves their connection weights to find a suitable level of complexity. The techniques that facilitate evolving a population of diverse and increasingly complex networks are described in detail in Stanley and Miikkulainen [9]; Stanley and Miikkulainen [10]; the important concept for the approach in this paper is that NEAT implements a solution to the *competing conventions problem* (a.k.a. the Permutation problem) through *historical markings* thereby enabling crossover and speciation for neural network genomes.

The NEAT implementation for this paper differs in a few details. First, crossover is on a neuron basis rather than connection basis, that is, genes are selected from parents through matching neurons. Thus the unit of crossover is a neuron rather than a connection. This change is inspired by the idea of neurons as feature extractors and detectors [14]. Second, this change to a neuron-based crossover change the way compatibility is counted. Excess connections are now defined as mis-matching connections on matched neurons. Disjoint connections are defined connections on mis-matching neurons. In addition, the count of mis-matching neurons now is incorporated into the compatibility metric. Third, crossover and mutation are mutually exclusive, that is, a new genome is either created through crossover without mutation or cloning with mutation, but mutation is never applied after crossover. Finally, the number of new species allowed to be created each generation is limited to one.

# 3 Speciation Heuristics

This paper investigates four different genotypic-based speciation heuristics. First is First Compatible, the original heuristic in NEAT, wherein genomes are placed with the first species that has compatibility below the current threshold (Algorithm 1). If no species is below the threshold, either a new species is created for the genome (if no new species has already been created) or the genome is placed with the most compatible of the species.

Second, Most Compatible places genomes with the species that is the most compatible (Algorithm 2). If no species is below the threshold, either a new species is created for the genome (if no new species has already been created) or the genome is placed with the most compatible of the species. This heuristic is a logical extension of First Compatible because genomes would ideally always be matched with their most compatible counterparts.

Third, Parental looks *only* at the species of the genome's parents to determine compatibility (Algorithm 3). If the none of the parent species are below the threshold, either a new species is created for the genome (if no new species has already been created) or the genome is placed with the most compatible of the parent and new species. This approach is intuited from nature, in that a child belongs either to its parent species or to a new sub-species of the parent species.

Finally, the fourth is Uncanny Valley looks first at the species of the genome's parents (Algorithm 4). If the genome is below the threshold for the parent, then the remaining species are investigated to see if there is another, non-parent species below the threshold. If there exists an additional compatible species (that is not the parent), the genome is placed in that species, otherwise the genome is placed with the parent species. If the genome is not compatible with

---

**Input**: Genome to be speciated
**Output**: Species to which the genome is assigned
Selected Species = **null**;
Minimum Compatibility = ∞;
**foreach** *Species s in current set of species* **do**
    Compatibility = GetCompatibility(s, Genome);
    **if** *Compatibility < Minimum Compatibility* **then**
        Minimum Compatibility = Compatibility;
        Selected Species = **s**;
    **if** *Compatibility ≤ Threshold* **then**
        **break**;
**if** *Minimum Compatibility > Threshold* **then**
    **if** *New Species == null* **then**
        New Species = CreateNewSpecies(Genome);
        Selected Species = New Species;
        Add New Species to current set of species;
Output Selected Species;

---

**Algorithm 1.** First Compatible Speciation Heuristic

---

**Input**: Genome to be speciated
**Output**: Species to which the genome is assigned
Selected Species = **null**;
Minimum Compatibility = ∞;
**foreach** *Species s in current set of species* **do**
    Compatibility = GetCompatibility(s, Genome);
    **if** *Compatibility < Minimum Compatibility* **then**
        Minimum Compatibility = Compatibility;
        Selected Species = **s**;
**if** *Minimum Compatibility ¿ Threshold* **then**
    **if** *New Species == null* **then**
        New Species = CreateNewSpecies(Genome);
        Selected Species = New Species;
        Add New Species to current set of species;
Output Selected Species;

---

**Algorithm 2.** Most Compatible Speciation Heuristic

its parent species, either a new species is created for the genome (if no new species has already been created) or it is placed with the most compatible of the parent and new species. This heuristic is named after the "uncanny valley" principle from robotics [7], since it follows a similar pattern where there is a dip in probability of being placed with your parents the closer you are in compatibility. This heuristic is similar to "anti-incest" approaches [5] in that it suppresses mating between substantially similar genomes by encouraging genomes to join the next most similar species. The next section details the experiments that reveal the differences resulting from these speciation heuristics.

---

**Input**: Genome to be speciated
**Output**: Species to which the genome is assigned
Selected Species = **null**;
Minimum Compatibility = ∞;
**foreach** *Species s in species of genome's parents* **do**
    Compatibility = GetCompatibility(s, Genome);
    **if** *Compatibility < Minimum Compatibility* **then**
        Minimum Compatibility = Compatibility;
        Selected Species = **s**;
**if** *Minimum Compatibility > Threshold* **then**
    **if** *New Species == **null*** **then**
        New Species = CreateNewSpecies(Genome);
        Selected Species = New Species;
    **else**
        Compatibility = GetCompatibility(New Species, Genome);
        **if** *Compatibility < Minimum Compatibility* **then**
            Selected Species = New Species;
Output Selected Species;

---

**Algorithm 3.** Parental Speciation Heuristic

## 4   Experimental Approach

Each of the speciation heuristics are tested in two domains under 0.1 and 0.9 crossover rates to demonstrate the effects that speciation heuristic selection has on performance and crossover. First is the XOR problem, an important calibration domain to ensure neuro-evolution can correctly solve and optimize nonlinear functions. XOR is a logical operator that returns true iff only one of the inputs is true. True is represented by +1 and false is represented by −1. The two inputs to XOR must be combined at a hidden unit, as opposed to only at the output. In the experiments with XOR, evolution is limited to 1000 generations and will not terminate at the first solution.

The second domain examined is the double-pole balancing domain [10], which is a well-known benchmark in reinforcement learning domains. In double-pole balancing, two poles are attached at a hinge to a movable cart. The role of the learning agent is to discover how to keep both pole elevated and not allow either to hit the cart. The agent keeps the poles elevated by instructing the cart to move at particular velocity and within the boundaries of the track. Fitness is determined by the number of time-steps the agent keeps the poles elevated, which is capped at 100000 for these experiments. Furthermore, the number of generations is limited to 500 and evaluation of a given run is stopped once a solution that achieves 100000 time steps is discovered. The next section describes the results of these experiments with speciation and crossover.

---

**Input**: Genome to be speciated
**Output**: Species to which the genome is assigned
Selected Species = **null**;
Minimum Compatibility = ∞;
**foreach** *Species s in species of genome's parents* **do**
  Compatibility = GetCompatibility(s, Genome);
  **if** *Compatibility < Minimum Compatibility* **then**
    Minimum Compatibility = Compatibility;
    Selected Species = **s**;
**if** *Minimum Compatability < Threshold* **then**
  **foreach** *Species s in current set of species* **do**
    Compatibility = GetCompatibility(s, Genome);
    **if** *Compatibility < Threshold* **then**
      Minimum Compatibility = Compatibility;
      Selected Species = **s**;
**else**
  **if** *New Species == null* **then**
    New Species = CreateNewSpecies(Genome);
    Selected Species = New Species;
  **else**
    Compatibility = GetCompatibility(New Species, Genome);
    **if** *Compatibility < Minimum Compatibility* **then**
      Selected Species = New Species;
Output Selected Species;

---

**Algorithm 4.** Uncanny Valley Speciation Heuristic

## 5   Results

All results are averaged over 40 runs and under identical settings (Appendix A) except for speciation heuristic and crossover rate. In the XOR domain with a low crossover (high asexual) rate of 10 %, all the speciation heuristics find a solution and optimizes to a perfect solution of XOR within 1000 generations (Fig. 1a), but demonstrate different learning trajectories getting there. While Most Compatible and Parental are substantially similar, the Uncanny Valley quickly spike upward in fitness and the Most Compatible slowly reaches the optimal. Figure 2a show the number of generations to achieve the first (not most exact) solution to XOR. The Uncanny Valley finds a solution to XOR significantly ($p < 0.01$) faster than the other speciation heuristics at an average of 47.6 generations to the first solution. The First Compatible and Parental are the next faster performers and are not significantly different with 97.2 and 122.5 generations on average, respectively. Finally, Most Compatible is significantly ($p < 0.01$) worse than the other heuristics at 445.8 generations on average.

In contrast, a high crossover rate of 90 % results in First Compatible (874.6 generations to first solution), Most Compatible (987.6 generations), and Parental (857.9 generations) as having similarly dismal performances (Fig. 1b) and, in

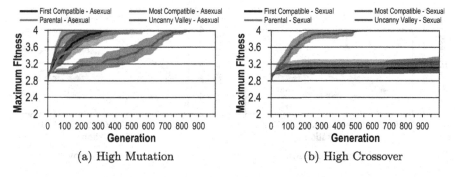

**Fig. 1.** Maximum fitness for XOR by speciation approach. The maximum fitness (averaged over 40 runs) for each generation is shown, surrounded by a 99 % confidence interval, under high mutation (a) and high crossover (b) rates.

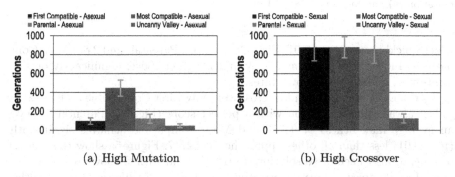

**Fig. 2.** Average generations to first solution for XOR by speciation approach. The generation count to the first genome that solves XOR (averaged over 40 runs) is shown with a 99 % confidence interval for high mutation (a) and high crossover (b) rates.

fact, do not find a solution to XOR in a majority of runs (Fig. 2b). However, Uncanny Valley (123.7 generations) significantly ($p < 0.01$) outperforms the other approaches at all generations past generation 50. Indeed, the time to first solution for Uncanny Valley with high crossover is not significantly different from First Compatible and Parental with low crossover rates.

Further contrast in the differences in heuristic and the effect of crossover can be seen in the average population fitness (Fig. 3). Similar to the maximum fitness results for the high asexual reproduction, the average population fitness under high asexual reproduction results in the Most Compatible and Parental as being statistically the same at most generations and the Uncanny Valley and Most Compatible as different from all other heuristics with significance $p < 0.01$. Interestingly, the final average population fitness scores for First Compatible (3.2) and Parental (3.2) are higher than Uncanny Valley (2.9), though Uncanny Valley's max fitness is better, while Most Compatible (2.7) remains the worst performer. For the high crossover rate setting, the average population fitness of Uncanny Valley (3.6) is significantly ($p < 0.01$) greater than the other speciation heuristics,

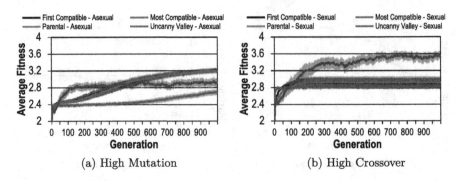

(a) High Mutation                    (b) High Crossover

**Fig. 3.** Average population fitness for XOR by speciation approach. The mean population fitness (averaged over 40 runs) is shown for each of the 1000 generations and surrounded with a shaded region representing a 99 % confidence interval under high mutation (a) and high crossover (b) rates.

which quickly converge to 2.91 (First Compatible, Parental) and 2.83 (Most Compatible). Indeed, the Uncanny Valley under high crossover significantly outperforms itself under low crossover rates.

In the DPB domain with low crossover, only First Compatible and Uncanny Valley always find a solution with a perfect score (Fig. 4a), Parental achieves an average max fitness of 95716 and Most Compatible achieves significantly ($p < 0.01$) less than all other approaches at 82717. Figure 5a show the number of generations to achieve a solution. In this domain with the low crossover rate, only Most Compatible (148.8 generations) is significantly different from the other heuristics. Most Compatible performs worse than Parental (53.1), First Compatible (34.6), and Uncanny Valley (19.2).

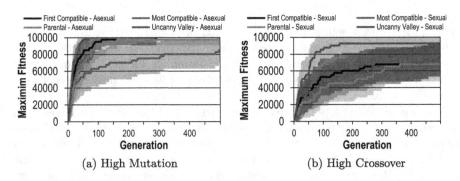

(a) High Mutation                    (b) High Crossover

**Fig. 4.** Maximum fitness for Double Pole Balancing (DPB) by speciation approach. The maximum fitness (averaged over 40 runs) is shown for each generation until the stopping criteria is met (performance of 100000 time steps or up to 500 generations) and surrounded with a shaded region representing a 99 % confidence interval under high mutation (a) and high crossover (b) rates.

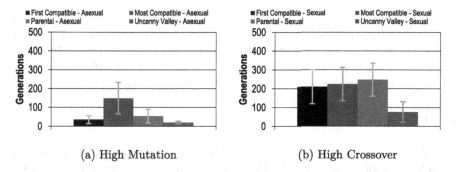

(a) High Mutation                    (b) High Crossover

**Fig. 5.** Average generations to a solution for Double Pole Balancing (DPB) by speciation approach. The generation count to the first genome that meets the stopping criteria (performance equal to 100000 time steps) for DPB (averaged over 40 runs) is shown with a 99 % confidence interval for high mutation (a) and high crossover (b).

Under a high crossover rate, none of the heuristics find a solution in every run within 500 generations. However, Uncanny Valley achieves an average champion fitness of 92647 that significantly ($p < 0.01$) outperforms the non-significantly different First Compatible (72590), Most Compatible (72678), and Parental (67756) at all generations past 70 (Fig. 4b). Indeed, the generations to a solution (Fig. 5b) for Uncanny Valley (75.3) is significantly fewer than First Compatible (210.1), Most Compatible (224.2), and Parental (248.1).

The average population fitness (Fig. 6) for DPB provides further insight into the interaction of speciation and crossover. In the low crossover scenario, there is no significant difference among the average population fitness by speciation heuristic, through the Uncanny Valley and First Compatible terminate early due to finding solutions. However, the high crossover scenario reveals differences

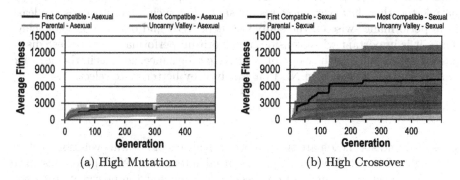

(a) High Mutation                    (b) High Crossover

**Fig. 6.** Average population fitness for Double Pole Balancing (DPB) by speciation approach. The mean population fitness (averaged over 40 runs) is shown for each generation or until the stopping criteria of a champion with a 100000 time step performance is found and surrounded with a shaded region representing a 99 % confidence interval under high mutation (a) and high crossover (b).

among the approaches. The final average population fitness scores for First Compatible (7189) is significantly ($p < 0.01$) higher than the other heuristics. Interestingly, Uncanny Valley's average population fitness reverses the pattern from its max fitness, that is, Uncanny Valley's average population fitness of 1772 is the lowest of the heuristics, while its max fitness is the highest.

## 6    Discussion

In nature, crossover is the primary (or only) form of reproduction for the most complex of organisms, especially those whose intelligence neuro-evolution (NE) is attempting to replicate. However, crossover in NE has not been beneficial, resulting in degraded performance in approaches that rely in crossover. Indeed, this paper provides evidence that crossover negatively affects NE performance.

Speciation has become an important aspect of NE approaches, with significant research examining measures to differentiate genomes into species to preserve diversity. This paper showed that different heuristics to place genomes into species based off a particular metric can enhance or impede performance. The Uncanny Valley speciation heuristic showed benefits in both the XOR and DPB domains, significantly improving performance over the baseline NEAT heuristic (First Compatible) as well as two other intuitively sensible heuristics (Most Compatible and Parental). Speciation improves performance of crossover by acting as a means of limiting the pool of candidates that may mate. Thus speciation, appropriately implemented in artificial evolution, can provide benefits to crossover by maintaining appropriate candidates pools.

Note, this paper doesn't claim one speciation heuristic is better than another, rather the deeper concept is that seemingly small implementation details (e.g. the speciation heuristic) can have a significant effect on the performance of an algorithm and interaction effects with varied parameters. Indeed, one naive implementation of speciation in NEAT is to group genomes that are most compatible together; however, these results show that such an approach is the least ideal performance-wise on the simple tasks of XOR and double pole balancing. Thus, it is important when considering algorithm performance that these disregarded details may be the cause of negative performance, such as in the negative performance of crossover in NE that is solved by better mate selection.

## 7    Conclusion

Crossover and speciation are two important features of natural evolution in their own right that can have significant mutual interaction. NE approaches have embraced speciation, but have not been able to exploit the power of crossover. Indeed, experiments in this paper demonstrated that crossover can be detrimental to performance in NE. However, results shows that different speciation heuristics change the performance of crossover. In fact, careful selection of speciation heuristic can allow NE to perform as well under high crossover as low crossover and improve performance under all conditions. Overall, speciation plays an important role in NE in preserving diversity, but a well-designed speciation heuristic has the potential to improve performance and unlock the power of crossover.

**Acknowledgments.** This work was supported and funded by the SSC Pacific Naval Innovative Science and Engineering (NISE) Program.

## A    NEAT Parameters Shared Across Experiments

| Parameter name | Parameter value |
|---|---|
| Population size | 200 |
| Minimum species # | 5 |
| Maximum species # | 15 |
| Interspecies mating rate | 0.01 |
| Selection proportion | 0.5 |
| Elitism proportion | 0.01 |
| Disjoint weight coefficient | 1 |
| Excess weight coefficient | 1 |
| Matching weight coefficient | 0.4 |
| Recurrence | no |
| Maximum weight magnitude | 5 |
| Initial connection ratio | 1 |
| Probability mutate weights | 0.9 |
| Probability add neuron | 0.01 |
| Probability add connection | 0.08 |
| Probability delete neuron | 0.001 |
| Probability delete connection | 0.09 |
| Probability select fitter gene | 0.5 |
| Probability recombine excess (fitter genome) | 0.8 |
| Probability recombine excess (less fit genome) | 0.15 |

## References

1. Barton, N.H., Bengtsson, B.O.: The barrier to genetic exchange between hybridising populations. Heredity **57**(3), 357–376 (1986)
2. Bernstein, H., Byerly, H.C., Hopf, F.A., Michod, R.E.: Sex and the emergence of species. J. Theor. Biol. **117**(4), 665–690 (1985)

3. Colegrave, N.: Sex releases the speed limit on evolution. Nature **420**(6916), 664–666 (2002)
4. Deb, K., Goldberg, D.E.: An investigation of niche and species formation in genetic function optimization. In: Proceedings of the 3rd International Conference on Genetic Algorithms, pp. 42–50. Morgan Kaufmann Publishers Inc. (1989)
5. Eshelman, L.J., Schaffer, J.D.: Preventing premature convergence in genetic algorithms by preventing incest. ICGA **91**, 115–122 (1991)
6. Mayr, E.: The Growth of Biological Thought: Diversity, Evolution, and Inheritance. Harvard University Press, Cambridge (1982)
7. Mori, M.: Bukimi no tani [the uncanny valley]. Energy **7**(4), 33–35 (1970). (online)
8. Mouret, J.-B., Doncieux, S.: Encouraging behavioral diversity in evolutionary robotics: an empirical study. Evol. Comput. **20**(1), 91–133 (2012)
9. Stanley, K.O., Miikkulainen, R.: Evolving neural networks through augmenting topologies. Evol. Comput. **10**, 99–127 (2002)
10. Stanley, K.O., Miikkulainen, R.: Competitive coevolution through evolutionary complexification. J. Artif. Intell. Res. **21**, 63–100 (2004)
11. Taylor, M.E., Whiteson, S., Stone, P.: Comparing evolutionary and temporal difference methods in a reinforcement learning domain. In: Proceedings of the Genetic and Evolutionary Computation Conference (GECCO 2006), pp. 1321–1328. ACM Press, New York, July 2006
12. Whiteson, S.: Improving reinforcement learning function approximators via neuroevolution. In: AAMAS 2005, Proceedings of the Fourth International Joint Conference on Autonomous Agents and Multiagent Systems, pp. 1386–1386. ACM, New York (2005)
13. Whiteson, S., Whiteson, D.: Stochastic optimization for collision selection in high energy physics. In: IAAI 2007, Proceedings of the Nineteenth Annual Innovative Applications of Artificial Intelligence Conference, Vancouver, British Columbia, Canada. AAAI Press, July 2007
14. Yao, X.: Evolving artificial neural networks. Proc. IEEE **87**(9), 1423–1447 (1999)

# Author Index

Printed in the United States
By Bookmasters